Bio-Based Polymers

Supervisor:

Yoshiharu Kimura

Copyright © Y. Kimura, 2013
All rights reserved. No part of this book may be reproduced in any form, by photostat, microfilm, retrieval system or any other means, without the written permission of authors.

ISBN978-4-7813-0271-3 C3058 ¥30000E
Printed in Japan

CMC Publishing Co., Ltd.
1-13-1 Uchikanda, Chiyoda-ku, Tokyo   101-0047 JAPAN

Preface

Synthesis of polymeric materials from renewable natural resources dates back only to the beginning of this century. In spite of this short history, these polymeric materials have now been accepted as "biobased materials" or more specifically "biobased polymers". This prompt acceptance of these new materials is based on the belief that their use efficiently contributes to suppressing the increase in carbon dioxide content in the global atmosphere, which is generally explained by the character of "carbon neutral" or "carbon offset". It is also believed that the development of biobased materials may possibly open a reliable route to new bio-industries that should be superior to the current oil-based industries in terms of sustainability. Until now, a considerable number of biobased polymers have already been proposed, and some of them are really industrialized. They are generally synthesized from naturally occurring biomass resources by the combination of chemical and biological technologies, which is admitted as "white biotechnology" or "industrial biotechnology". For replacing the ordinary oil-based plastics including engineering plastics, the bio-based polymers ought to have excellent functional properties and high performance. Currently, such high-performance biobased polymers, both fully and partly biobased ones, have already been developed. These novel polymers can be combined with the conventional plastic materials to create a new polymer platform in the polymer science and engineering of the future.

This book is dedicated to the research and development communities in academia and in industry to promote the industrialization of the biobased polymers mentioned above and to support the teaching of the advances in new materials in universities and other societies. The contents may not be comprehensive because naturally occurring polymers are not included in order to concentrate on the newly developed biobased polymers of synthetic origin. However, new trials for developing biobased polymers are mostly covered, and each chapter of this book is full of new information and idea, being really useful not only for scientists and engineers of materials fields but also for other experts of different disciplines.

Publishing of this book was achieved by choosing authors of the individual chapters on the basis of their expertise and their excellent contributions to the research fields. I'm very grateful to these scientists for their willingness and engagement in contributing their book chapters. Without their effort the publication would not have been done. I don't know how I can apologize for the delayed publication of this book. The main reason was that I myself had not submitted my manuscript until now. I deeply regret my busy situation for the last several years. Under these circumstances, publication of this book has been accomplished by virtue of the special aid of the publisher CMC Publishing Co. Ltd. Here, I express my greatest thanks to CMC Publishing for their patience as well as their effort in editing and publishing this book in excellent quality. Special thanks are due to Ms Takako KADOWAKI and her CMC Publishing colleagues for their constant suggestions and kind supports.

February 21, 2013
Yoshiharu Kimura

List of Contributors

Yoshiharu Kimura	Professor, Kyoto Institute of Technology, Department of Biobased Materials Science
Hajime Nakajima	Post-doctoral researcher, Kyoto Institute of Technology
Maria José Climent	Professor, Instituto de Tecnología Química (Universidad Politécnica de Valencia-Consejo Superior de Investigaciones Científicas)
Maria Mifsud	Professor, Instituto de Tecnología Química (Universidad Politécnica de Valencia-Consejo Superior de Investigaciones Científicas)
Sara Iborra	Professor, Instituto de Tecnología Química (Universidad Politécnica de Valencia-Consejo Superior de Investigaciones Científicas)
Hitomi Ohara	Professor, Kyoto Institute of Technology, Department of Biobased Materials Science
Hideki Yamane	Professor, Kyoto Institute of Technology, Department of Biobased Materials Science
Tadahisa Iwata	Professor, The University of Tokyo, Department of Biomaterial Sciences, Graduate School of Agricultural and Life Sciences
Takeharu Tsuge	Associate Professor, Tokyo Institute of Technology, Department of Innovative and Engineered Materials
Sei-ichi Taguchi	Professor, Hokkaido University, Graduate School of Engineering
Hideki Abe	Team Reader, RIKEN Biomass Engineering Program, Bioplastic Research Team
Toshihisa Tanaka	Associate Professor, Shinshu University, Faculty of Textile Science and Technology
Mureo Kaku	Manager, DuPont Kabushiki Kaisha, Industrial Biosciences
Kotaro Satoh	Associate Professor, Nagoya University, Department of Applied Chemistry, Graduate School of Engineering
Masami Kamigaito	Professor, Nagoya University, Department of Applied Chemistry, Graduate School of Engineering
Sei-ichi Aiba	Senior Research Scientist, National Institute of Advanced Industrial Science and Technology (AIST), Biological Substance Engineering Research Group, Bioproduction Research Institute
Shiro Kobayashi	Distinguished Professor, Kyoto Institute of Technology, Center for Nanomaterials and Devices
Masatsugu Mochizuki	Professor, Kyoto Institute of Technology, Center for Fiber and Textile Science

Contents

Chapter 1   General Introduction: Overview of the current development of biobased polymers
Hajime Nakajima, Yoshiharu Kimura

1. 1	Concept of biobased polymers	1
1. 2	Biobased polymers and biodegradable polymers	1
1. 3	Current biobased polymers	2
1. 3. 1	Biomass polymers	2
1. 3. 2	Bio-engineered polymers	3
1. 3. 3	New metabolite polymers from bio-originated building blocks	3
1. 3. 4	Conventional petrochemical polymers from bio-derived monomers	4
1. 4	New biobased polymers	4
1. 5	High-performance biobased polymers	4
1. 6	Growing production of biobased polymers	6
1. 7	Biomass refinery	7
1. 7. 1	Biobased building blocks from cellulose	9
1. 7. 2	Production of bio-succinic acid	9
1. 7. 3	1.3-Propanediol and 3-hydroxypropionic acid from glycerol	9
1. 8	Industrialization of PLLA	10
1. 8. 1	NatureWorks LLC	10
1. 8. 2	PURAC	11
1. 8. 3	PURAC-Sulzer Chemtech-Synbra	12
1. 8. 4	PURAC-Arkema	13
1. 8. 5	PURAC-Indorama	13
1. 9	Industrialization of sc-PLA	13
1. 9. 1	FKuR/Synbra	14
1. 9. 2	Teijin	14
1. 10	Manufacturing of bacterial polyesters	14
1. 11	Biodegradable poly(butylene succinate) and its copolymers	15
1. 12	Other biobased polymers	15
1. 12. 1	Furanics and biobased terephthalic acid	15
1. 12. 2	Various biobased polyamides	16
1. 13	Latest examples of high-performance biobased polymers	18
1. 13. 1	Bio-based LCP	18

1. 13. 2	Isosorbide-containing polymers	18
1. 13. 3	Terpene-derived polymers	19
1. 13. 4	Specialty PLA polymers from modified lactides	20
1. 14	New platform of biobased polymers	21

Chapter 2　Biomass and Biomass Refining
Maria José Climent, Maria Mifsud and Sara Iborra

2. 1	Introduction	25
2. 2	Carboxylic acids and polyols	26
2. 2. 1	Lactic acid	26
2. 2. 2	Succinic acid	28
2. 2. 3	3-Hydroxypropionic acid	30
2. 2. 4	Levulinic acid	31
2. 2. 5	Glycerol	32
2. 2. 6	Furans	35
2. 2. 7	5-Hydroxymethyl furfural (HMF)	36
2. 2. 8	Furfural	38

Chapter 3　Bio-polyesters

3. 1	Poly(lactic acid) ················ Hitomi Ohara	45
3. 1. 1	Lactic acid fermentation	45
3. 1. 2	Synthesis of poly(lactic acid)	49
3. 1. 3	Industrial manufacturing methods	55
3. 1. 4	Direct polycondensation	57
3. 1. 5	Conclusion	58
3. 2	Stereocomplex PLA ················ Hideki Yamane	61
3. 2. 1	Introduction	61
3. 2. 2	Melt-blending of PLLA and PDLA	62
3. 2. 3	Melt-spinning of PLLA/PDLA blend	64
3. 2. 4	Biaxially oriented PLLA/PDLA blend films	67
3. 2. 5	Conclusion	69
3. 3	Polyhydroxyalkanoate ················ Tadahisa Iwata, Takeharu Tsuge, Sei-ichi Taguchi, Hideki Abe and Toshihisa Tanaka	71
3. 3. 1	General introduction	71
3. 3. 2	Biosynthesis of P(3HB) and its copolymers	72
3. 3. 3	Fermentative production and mechanical properties of PHB and its copolymers	73
3. 3. 4	Ultra-high-molecular-weight P(3HB)	75
3. 3. 5	Structure of P(3HB)	77
3. 3. 6	Fibers of P(3HB) and its copolymer	77
3. 3. 7	UHMW-P(3HB) fibers	78
3. 3. 8	Structure and function of PHB depolymerase	79
3. 3. 9	Industrial production of P(3HB) and its copolymers	80
3. 4	Poly (trimethylene terephthalate, PTT) ················ Mureo Kaku	86
3. 4. 1	Introduction	86
3. 4. 2	Bio-1,3 propanediol (Bio-PDO^{TM})	87

3. 4. 3	Poly (trimethylene terephthalate, PTT)	88
3. 4. 4	Sorona® polymer for fiber applications	88
3. 4. 5	Sorona® polymer for injection mold applications	91
3. 4. 6	Summary	93

Chapter 4   New Polymerization Methods for Bio-based Polymers

4. 1   New polymerization methods for bio-based polymers from renewable vinyl monomers ············ **Kotaro Satoh and Masami Kamigaito** ···· 95
- 4. 1. 1   Introduction ··· 95
- 4. 1. 2   Controlled/Living polymerization of petrochemical vinyl monomers ··· 96
- 4. 1. 3   Polymerizations of naturally-occurring olefins (Terpenes) ··· 98
- 4. 1. 4   Polymerizations of naturally-occurring styrenes (phenylpropanoids) ··· 103
- 4. 1. 5   Naturally-derived acrylic monomers ··· 107
- 4. 1. 6   Conclusion ··· 108

4. 2   Biobased polyamides ············ **Sei-ichi Aiba** ···· 112
- 4. 2. 1   Introduction ··· 112
- 4. 2. 2   Biobased PAs in the market and under R & D ··· 113
- 4. 2. 3   Conclusion ··· 118

4. 3   Enzymatic polymerization catalyzed by hydrolases ············ **Shiro Kobayashi** ···· 120
- 4. 3. 1   Introduction ··· 120
- 4. 3. 2   Characteristics of enzymatic reactions and basic concept of enzymatic polymerization ··· 120
- 4. 3. 3   Synthesis of polysaccharides ··· 123
- 4. 3. 4   Synthesis of polyesters ··· 134
- 4. 3. 5   Conclusions and future perspectives ··· 158

Chapter 5   Application of Bio-based Polymers
Masatsugu Mochizuki

- 5. 1   Introduction ··· 165
- 5. 2   Key performance features of PLA ··· 166
  - 5. 2. 1   Chemical, physical and thermal properties ··· 166
  - 5. 2. 2   Biodegradability and their biodegradation mechanism ··· 166
  - 5. 2. 3   Environmental sustainability ··· 167
- 5. 3   Processing of PLA ··· 167
  - 5. 3. 1   Melt crystallization and cold crystallization ··· 167
  - 5. 3. 2   Crystallization rate of PLA ··· 168
- 5. 4   High-performance PLA ··· 169
  - 5. 4. 1   Heat resistance ··· 170
  - 5. 4. 2   Hydrolysis resistance ··· 170
  - 5. 4. 3   Impact strength ··· 170
- 5. 5   PLA products and potential applications ··· 171
  - 5. 5. 1   Fibers and nonwovens ··· 171
  - 5. 5. 2   Films and sheets ··· 172
  - 5. 5. 3   Injection molding ··· 172
  - 5. 5. 4   Thermoforming ··· 173
  - 5. 5. 5   Foaming molding ··· 173

Chapter 1
General Introduction:
Overview of the current development of biobased polymers

Hajime Nakajima, Yoshiharu Kimura

1.1 Concept of biobased polymers

With recognition of the constraints of natural resources and of the serious climate change due to the pollution of global environment, much effort has been paid to replace the energy and material sources from fossil resources to renewable natural feedstock. In materials fields, the development of renewable materials by using biomass resources has attracted much attention of scientists and engineers of different fields. These new materials are currently accepted as biobased materials. Since the carbon atoms contained in the biomass-derived materials are originated from the atmospheric carbon dioxide (fixed through the photosynthesis of plants), incineration of these materials would not give additional carbon dioxide to the global environment. Therefore, their use is believed to be superior to the current use of the oil-based materials in suppressing the increase of carbon dioxide content in the global atmosphere. Until now, several polymeric materials developed with this new concept have been manufactured by industries as "biobased polymers" [1,2]. They are generally synthesized from naturally occurring polysaccharides or their breakdown substances by the combination of chemical and biological technologies, which is admitted as white bio-technology.

1.2 Biobased polymers and biodegradable polymers

Many of the biobased polymers developed thus far were once utilized as bioabsorbable polymers for medical use [3] and as biodegradable plastics for waste management [4]. Table 1 compares the bioabsorbable, biodegradable, and biobased polymers that can be classified in the historical order. Even now, the first generation bioabsorbable polymers are particularly important for use as cell-matrices in regenerative medicine, whereas the second generation biodegradable polymers are essential in the application to packaging and agricultural mulching. In the beginning of this century a few biodegradable polymers such as poly-L-lactide (PLLA), poly(3-hydroxyalkanoate) (PHA), poly(butylene succinate) (PBS), etc. were manufactured. However, their production sizes

Table 1 Current status of biodegradable/biobased polymers and their application

	First Generation Bioabsorbable polymers	Second Generation Biodegradable polymers	Third Generation Biobased polymers
Character	Bioabsorbable/ degradable	Biodegradable/eco-friendly	Biobased, derived from renewable resources
Purpose	Temporal replace of bio-tissues	Replace of commodity plastics for waste management	Replace of structural materials
Applications	Hydrolysis rate biomedical materials (DDS, bone fixation, sutures)	Short life daily appliances garbage bags	Long life electric appliances automobile parts
Examples	Poly(α-hydroxy acids) PGA, PLLA, Peptide, etc.	Aliphatic polyesters PLLA (NatureWorks®) PHB, PBS (GS Pla®) PBSA (Ecoflex®) PEAT (Apexa®), etc.	Various polymers PLLA, sc-PLA, PTT, PMBL, Bio-succinates, etc.
Social Recognition	Tissue engineering	Legislative infra-structure	Environmental issue and brand strategy
Industrializaion	1970–80s	2000–	2008–

have remained small because of the limited market size of the biodegradable products. With this recognition, the alternative character of "bio-based" has become more important to forward the development of this class of polymers as the third generation polymers. Since these polymers can be synthesized from non-petroleum resources, they can be renewable and sustainable in terms of long-range production and consumption. In the development of bio-based polymers, substitution of a broader range of the conventional plastic materials has been considered than in the previous development of biodegradable polymers. For this reason, the biobased polymers ought to have high-performances and specialties together with higher stability and durability that should be comparable to those of the conventional polymeric materials.

1.3 Current biobased polymers

Table 2 shows the biobased polymers thus developed as well as under development [5]. These polymers are classified into two groups: one is of natural origin and the other is of synthetic origin. The first group of polymers of natural origin is subdivided into two groups based on whether the bio-production can be genetically controllable or not. On the other hand, the second group is sub-divided into two categories based on whether their synthesis is easily done with new biobased building blocks or with the conventional monomers of biomass origin, i.e., new biomass-based polymers are made of metabolites or petroleum-based polymers are made by using biobased techniques.

1.3.1 Biomass polymers

This sub-class involves natural polymers that are extracted from biomass feedstock and used

Table 2 Classification of biobased polymers

---
1) Natural polymers: naturally obtained
  Biomass polymers: direct use of biomass as polymeric materials including chemically modified ones
    Regenerated cellulose, Cellulose acetate, Starch-based materials, Chitin, Modified starch, etc.
  Bio-engineered polymers: bio-synthesized by using microorganisms and plants
    Poly(3-hydroxyalkanoate), Bacterial cellulose, Poly(glutamic acid), etc.
2) Synthetic polymers: synthesized from biomass-originated monomers
  New metabolite polymers from biomass feedstocks
    Polylactides (PLA), Poly(trimethylene terephthalate) (PTT), poly(ethylene furanoate) (PEF), Poly(butylene succinate) (PBS), Polytulipalin (PT), Polycarbonate (PC), Polyamides (Nylon), etc.
  Conventional petrochemical polymers from bio-derived monomers
    Bio-polyethylene (PE), Bio-polypropylene (PP), etc.

---

directly as polymeric materials. Typical examples are polysaccharides of natural origin. Their chemically modified derivatives are also included in this sub-class. Although the use of these polymers started in the late 19th century, it has not widely expanded, because of the difficulties in their isolation and purification. The strong intermolecular hydrogen bonds involved in these natural polymers make it difficult to control their morphology and physical properties although the biodegradable nature is retained as their original property.

1. 3. 2 Bio-engineered polymers

This sub-class involves polymers produced by microorganisms. Their production can be controlled by bio-technological methods using E. coli and plants, affording a new biological production process for polymers. In particular, bacterial celluloses and poly(3-hydroxyalkanoate)s (PHA) are bio-engineered by fermentation whose process can easily be controlled in industrial manner. Poly(3-hydroxybutanoate) (PHB) and its copolymers, known as the energy-storage materials accumulated in the cells of microorganisms [6], have been manufactured although their application has been rather limited until now.

1. 3. 3 New metabolite polymers from bio-originated building blocks

These polymers are all synthesized by polymerization of metabolic products that are obtained from naturally occurring substances by bio-refinery or break-down of biomass. This sub-group involves poly-L-lactide (PLLA) prepared from L-lactic acid, an aliphatic bio-polycarbonate (PC) prepared from isosorbide, polytulipalin (PT), and polyamides such as Nylon 11 prepared from an amino-carboxylic acid obtained from castor oil. The current poly(trimethylene terephthalate) (PTT) (Sorona® produced by Du Pont) [7] and poly(butylene succinate) (PBS) consisting of bio-derived 1,3-propanediol and bio-synthesized succinic acid, respectively, are only partly bio-based, because the other repeating units are petrochemically synthesized substances.

1. 3. 4 Conventional petrochemical polymers from bio-derived monomers

Many petrochemical products are produced from ethylene and propylene as the starting chemicals in the current petroleum chemistry. Bio-ethanol manufactured as an energy source can be converted to these starting chemicals and utilized to prepare polyethylene and other related polymers. At present, many attempts are made to synthesize poly(ethylene terephthalate) (PET) from ethylene glycol and terephthalic acid obtained by chemical conversion of bio-ethanol and bio-isobutanol [8].

1. 4 New biobased polymers

Currently, synthetic biobased polymers are developed by using two methods in conjunction with their classification mentioned above. One method is to newly develop biomass-based polymers from new bio-originated substances (metabolites) whilst the other is to make the conventional petroleum-based polymers from biomass-originated feedstock. The first generation biobased polymers comprised the former polymers as PLA and PHB for which cost performance and property control were not easy. In the second generation polymers, therefore, much effort has been made for replacing the monomers of petroleum-based polymers with bio-originated ones. Typical examples are ethylene and ethylene glycol synthesized from bio-ethanol and terephthalic acid from isobutanol. The other approach to this generation is to find monomers having similar structures with the conventional ones. In this case, 1,3-propanediol, some sort of diamines, and dicarboxylic acids are synthesized from biomass feedstock and used for making biobased polymer homologues showing similar properties. The current industries seem to prefer the second generation polymers because the existing facilities can be used for their manufacturing. The guaranteed properties and material life are also favorable for quick industrialization. With these conventional polymers, however, the material innovation may not be possible, and a new paradigm ought to be established in this century by developing new polymeric materials with the new concept of the first generation polymers.

Table 3 summarizes the new synthetic biobased polymers that have been developed or under development, including the partly bio-based polymers. Among them, PTT, PLLA, stereocomplex-type polylactides (sc-PLA), stereo-block polylactides (sb-PLA), PHA, and polyamide-11 are now manufactured, whereas aliphatic polyamides, polycarbonates consisting of isosorbide (derived from sorbitol), and poly(ethylene furanoate) (PEF) are under development. These polymers are obtained by polymerization of the monomers obtained by fermentation and other bio-refining processes from biomass.

1. 5 High-performance biobased polymers

Being developed within the platform of highly advanced petroleum industries, the new bio-based polymers ought to overcome the comparison with the conventional oil-based polymers with respect to the performance and cost prior to their large-scale industrialization. For their prompt marketing, therefore, it is necessary to impart distinct performance or functionality to the biobased polymers that cannot be attained with the conventional oil-based polymers [9]. Typical examples

Table 3 New biobased polymers synthesized from different building blocks

Class	Polymer	Monomer	Synthetic method	Developing	Characteristics
Polyolefins	PE	Ethylene	Chemical-coversion from bio-ethanol	Braskem	
	PP	Propylene			
Polyesters	PTT	1,3-Propanediol	Bio-conversion from glucose/glycerin	Du Pont, *Sorona*	Oil-based terephthalic acid
	PET	Ethylene glycol	Fermentation	Teijin, *Plantpet*	
		Terephthalic acid	Bio-conversion from isobutyl alcohol	Toray/Gevo	
	Copolyesters	Succinic acid	Fermentation	Du Pont, *Apexa* BASF, *Ecoflex*	PBS derivatives Partly biobased
		Terephthalic acid			
		Ethylene glycol	Fermentation		
		Butanediol	Chemical coversion from succinic acid		
	PLLA	L-Lactic acid	Fermentation	NatureWorks, *Ingeo* Purac	Massive
	sc-PLA	L-/D-Lactic acids	Fermentation	Teijin	Fibers & films
	sb-PLA	L-/D-Lactic acids	Fermentation	Musashino/Kimura	Neo-PLA
	PEF	Ethylene glycol	Fermentation	Avantium	
		2,5-Furandicalboxylic acid	From fructose		
	PBS	Succinic acid	Fermentation	Roquette/DSM Mitsubishi/Ajinomoto	
		1,4-Butanediol	Chemical coversion from succinic acid		
	PHA	Bacterial polyesters	Bacterial and plant synthesis	Metabolix, Kaneka	PHBV, PHBH Genetic modification
Polycarbonates	PC	Isosorbide	From sorbitol	Mitsubishi, *Durabio*	
Polyamides	Nylon 56	Adipic acid	×	Toray	
		Cardarin	Enzymatic decarboxylation of lysin		
	Nylon 4	γ-Aminobutyric acid	Enzymatic decarboxylation of glutamic acid	AIST	
	Nylon 610	Sebacic acid	Caster oil, licinoleic acid	BASF	Oil-based hexamethylenediamine
	Nylon 46	1,4-Diamionobutane	Fermentation	DSM	Oil-based adipic acid
	Nylon 11	11-aminoundecanoic acid	Caster oil	Arkema	
Acrylics	Butyrolactones	Tuliparin	Plant-based	Du Pont	
		MMT	From levulinic acid	Du Pont	
Polyurethanes	PU	Plant oil-based polyols	From caster oil	Du Pont, BASF	Oil-based diisocyanate
		Diisocyanate	×		

×: difficult to be synthesized by bio-processes
MMT: 3-methylene-5-methyltetrahydrofuran-2-one

of these biobased specialty polymers are acrylic polymers from tuliparin (extracted from tulip), 3-methylene-5-methylterahydrofuran-2-one (MMT: derived from levulinic acid), and an enol ester of pyruvic acid. A glassy lactide copolymer with mandelide (from mandelic acid) is another potentially biobased polymer with specialty. Sc-PLAs and sb-PLAs, consisting of enantiomeric lactide polymers, are also currently developed as high-performance polymers. Few of these trials, however, have been successful until now.

In the nearest future, high-performance and specialty biobased polymers will be manufactured with reasonable prices, replacing various petroleum-based polymers. In this introduction, the current status of synthetic biobased polymers are overviewed both from the industrial and scientific viewpoints.

1. 6   Growing production of biobased polymers

As described above, a certain size of market has already been established for biobased polymers, and the future market is believed to grow up promptly. However, it is still difficult to predict a precise market size in the near future, because the biobased polymers have found limited application until now, making it difficult to replace the conventional petroleum-derived polymers in large scale. The main reason for this inferiority is that the physical and chemical properties of the biobased polymers have not fully been controlled and that their specialties have rarely been established for competing with those of the conventional polymeric materials. Figure 1 shows a prediction on the growth of worldwide production of biobased polymers until 2020, which was reported in PRO-BIO 2009 [10]. This prediction was on the basis of the amounts that are scheduled by the manufacturers of the respective polymers. It is therefore trustworthy that the total production of biobased polymers will reach 2.33 and 3.45 Mt (million ton) in 2013 and 2020, respectively. According to this prediction, the annual market growth will be 37 % in between 2007 and 2013 and 6 % in between 2013 and 2020, rather slowing down in the coming decade. The most widely used biobased polymers in this period are starch-based polymers, poly(lactic

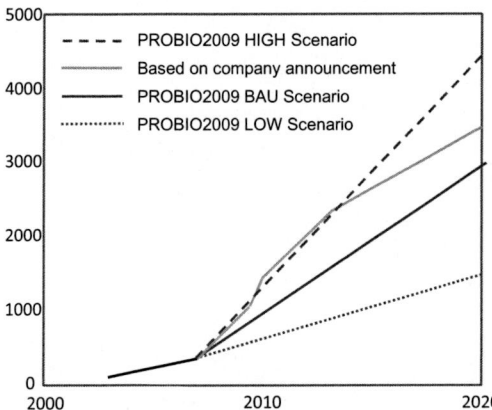

Figure 1   Prediction of the worldwide production capacity of biobased polymers until 2020 (ref: PROBIO2009)

acid) (PLA), biobased polyethylene (Bio-PE), and polyhydroxyalkanoates (PHAs). Their market sizes are estimated by taking the technical barriers, applications, costs, and easiness in supply of the raw materials into consideration. In the business-as-usual (BAU) scenarios, a stable growth is proposed in biobased epoxy resins in addition to the three of the four key polymers, i.e. starch-based plastics, PLA, and Bio-PE, while a modest growth is assumed in cellulose films, PHAs, and biobased polyurethane (PUR). The best scenario proposes a very strong growth of the above four polymers in addition to a stable growth of cellulose films, PHAs, and biobased PUR, making the total global production of biobased polymers reach 4.4 million Mt by 2020 which is approximately 30 % higher than that based on the above companies' announcements. This best scenario also includes a market of biobased polyamides (PA) and poly(trimethylene terephthalate) (PTT) in spite of their rather limited bulk application. The worst scenario proposed is based on a pessimistic progress both in the technologies and business conditions. Especially, little progress is expected in the developments of biobased succinic acid, biobased PA, and biobased polypropylene (PP). In general, the prediction of market growth for each of the specific products is easily influenced by whether the business condition may be improved or not in the future. A technical breakthrough will dramatically change the business situation and even affect the world economic state to lead the above optimistic prediction. It should be noted that in any of the scenarios a stable growth is promised in the biobased polymer market despite different growth rates predicted.

1. 7 Biomass refinery

The process for obtaining chemical substances from biomass feedstock is generally called bio-refinery in conjunction with the petroleum refinery conducted for fractionation and purification of crude oil. The technology for break-down of biomass is essential for establishing the biobased industries that produce and supply monomers and other chemicals. In 2004, the Research Institute of the Department of Energy (DOE) in the United States appointed 12 platform chemicals that can readily be obtained by biochemical treatment of biomass [11]. Figure 2 shows the structures of these platform chemicals. It is evident that eight species of the 12 chemicals are carboxylic acids, indicating that oxidation is favored rather than reduction in the bio-conversion or metabolism. Each of the platform chemicals was selected not only because of the easiness of synthesis but also because of their easy transformation to useful derivatives. Their main application is assumed to synthesize various polymeric materials by polycondensation and polyaddition.

On the other hand, several bio-originated substances such as lactic acid and trimethylene glycol were not appointed as the platform chemicals although they are important monomers for making PLA and PTT: biobased polymers that are now industrialized on a global scale. Also, sebacic acid, which is an important plant-derived monomer of aliphatic polyesters and polyamides, was not involved in the platform chemicals, either. It is probably because of the rather limited versatility of these substances in converting to other types of building blocks. Direct use of plant oils and fats as the building blocks is also possible. For example, partially hydrolyzed castor oil is used as a polyol for synthesizing bio polyurethane [12].

Table 4 summarizes the current status of development of bio-based monomers and chemicals

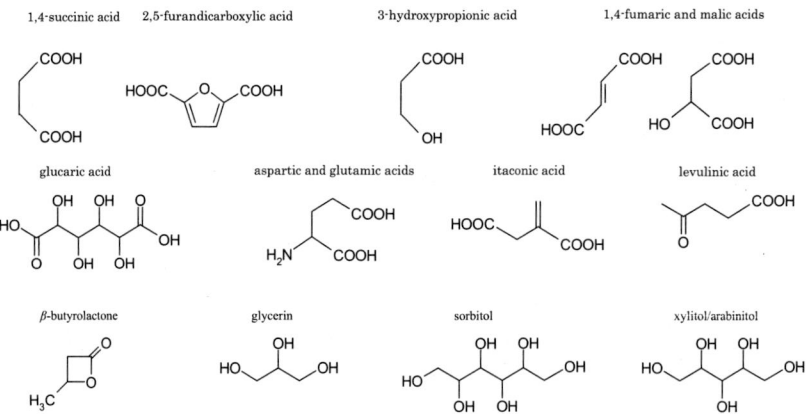

Figure 2 Platform chemicals targeted in the biorefining of biomass

Table 4 Comparison of the development stage of biobased monomers

carbon	chemical	Research stage	Development stage	Manufacturing stage
C2	ethanol (starch-based)			*
	ethanol (cellulose-based)		*	
	ethylene			*
C3	lactic acid			*
	1,3-propanediol			*
	acrylic acid		*	
C4	1,4-butanediol		*	
	1,4-diaminobutane	*		
	succinic acid			*
	isobutanol		*	
	fumalic acid	*		
C5+	itaconic acid			*
	furandicarboxylic acid		*	
	adipic acid		*	
	polyol			*
	terephthalic acid	*		

* Showing the development stage in a scale where the more matured takes the more rightward position.

in 2012 [13]. It is evident that the bio-synthesis of a number of key monomers such as succinic and itaconic acids has been progressing quite rapidly. Even terephthalic and adipic acids may possibly be replaced by the ones synthesized from biomass feedstock.

Whereas, the biomass feedstock produced by large-scale plantation is mostly used as energy source, and only a part of it is directed to fabrication of materials. Since most of the biomass is cropped in fibrous form, biomass itself can be utilized for reinforcement of plastic materials [14]. Recently, it has become possible to extract crystalline cellulose in nanofiber form from pulp. The cellulose nanofibers thus obtained are dispersed in the polymer matrix for strengthening. Since the modulus of these cellulose nanofibers reaches 140 GPa, which is four times the value of kenaf fibers and twice that of glass fibers, the resulting nanofiber/polymer composites exhibit excellent mechanical properties [15]. The composites are characterized by the various nano-effects of the dispersed fibers in addition to the high transparency preserved.

1. 7. 1 Biobased building blocks from cellulose

Glucose is the most convenient raw material for producing biobased chemicals, because it can readily be fermented into various substances that can be used as building blocks. Glucose is generally obtained from starch harvest or edible biomass feedstock. The mass use of starch by the industry, however, has been accused because it may possibly call for deficiency of grain crops in the food supply. For continuous growth of the biobased polymer industry, therefore, the glucose supply must be brought out of the supply chain of food. Additionally, recent grain price has risen up with increase in oil price, and non-starch feedstock must be sought for the cost-down of glucose feedstock. The most promising nonedible biomass feedstock is cellulosic biomass that is not only produced with grain crops but also harvested from various plant resources, although its breakdown to glucose is not easy [16]. Recently, Cobalt Technologies (USA) and American Process (USA) have announced an agreement to build the world first plant for industrialization of cellulosic 1-butanol. American Process also intends to open an ethanol plant which is based on utilization of the hemicellulosic waste coming out from wood paneling firms. Genomatica (USA) has also launched another cellulosic chemical production. Genomatica collaborates with an engineering firm Chemtex which is a subsidiary of Italian PET producer, Gruppo Mossi and Ghisolfi, to produce both ethanol and 1,4-butanediol. The chemical giant BASF (Germany) invests $ 30 million in a cellulosic sugar firm, Renmatrix, which has developed less expensive method for extraction of C_5 and C_6 sugars from cellulosic biomass using supercritical water. In spite of these activities, further development will be needed for mass-scale production of cellulosic glucose in order to compete with the established starch-based sugar production.

1. 7. 2 Production of bio-succinic acid

DSM (the Netherland) and Roquette (France) predict that the annual market of bio-succinic acid will be totally 2,000 kMt in 2020. (Figure 3(a)) [17]. Since succinic acid can be converted into a variety of C4 chemicals by using the conventional biological and chemical modifications (Figure 3(b)), bio-succinic acid may occupy a central position in the biobased industry. Its polymerization can give biobased polymers PBS and PBST commercialized as GS-Pla (Mitsubishi Chemical) and Apexa (Du Pont), respectively. There are many potential suppliers of bio-succinic acid, for example, BioAmber, Myriant, Joint Venture of DSM/Roquette, and Joint Venture of BASF/CSM.

1. 7. 3 1.3-Propanediol and 3-hydroxypropionic acid from glycerol

Natural fat, vegetable oil, and animal fats are also admitted as key biomass feedstocks because of their excellent cost performance, worldwide availability, and their numerous commercial applications themselves, especially in the bio-fuel field. Their basic chemical structures, triglycerides, are similar to those of the corresponding petroleum-derived analogues in terms of involvement of olefinic and paraffinic components. Hydrolysis of natural fats provides fatty acids whose

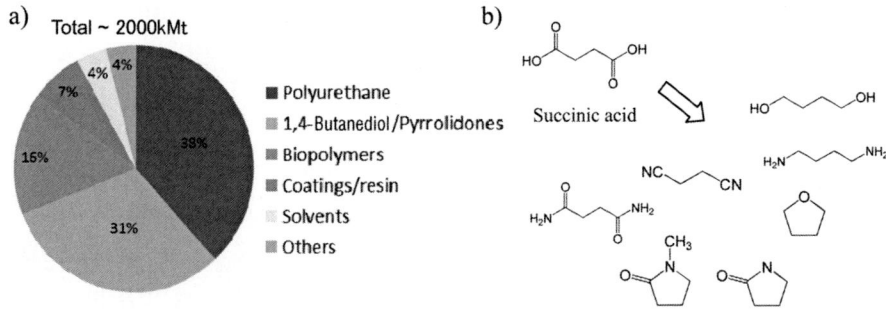

Figure 3 a) A prediction for the production of succinic acid and its derivatives in 2020 suggested by DSM and Roquette, b) Succinic acid and its derivatives via chemical and biological conversions suggested by DOE (USA)

methyl esters can be used as diesel oil in addition to glycerin. Since this glycerin has less utility, its derivatization to useful chemicals is focused with increase in biodiesel market. Recently, biochemical conversion of this abundant glycerin has become possible. Typical products of the bio-processes are 1,3-propanediol and 3-hydroxypropionic acid that are the key intermediates of several biobased polymers [18]. In particular, 3-hydroxypropionic acid can be a precursor of acrylic acid. It is also known that recinoleic acid obtained from caster oil can be a source of various biobased monomers such as sebacic acid, 11-aminoundecanoic acid, and 1,10-decanediamine [19].

1. 8 Industrialization of PLLA

Showing high mechanical properties together with biodegradable nature, PLLA was thought to provide a large opportunity to replace non-degradable oil-based polymers such as poly(ethylene terephthalate) (PET) and polystyrene (PS). In the early 1990s, therefore, Cargill Inc. (USA) started the production of poly-L-lactide (PLLA). They succeeded in synthesizing L-lactide in large scale and polymerizing high-molecular-weight PLLA by ring-opening polymerization (ROP) of L-lactide. Since then, PLLA has been utilized as biodegradable plastics for short-term applications such as rigid and flexible packaging films, cold drink cups, cutlery, filament and staple fibers, bottles, injection- and extrusion-molds, coatings, and so on, all of them being degraded under a controlled compositing condition [20]. Around 2000, the biobased nature of PLLA was highlighted, and its production as a biobased polymer started. This historical trace is well comparable to the three characters of PLLA polymers, i.e., bioabsorbable, biodegradable, and biobased. The current manufacturing status is briefly summarized as below.

1. 8. 1 NatureWorks LLC

The current world leader in manufacturing of PLLA is NatureWorks LLC (USA). The trade name of the PLLA is 'Ingeo®'. NatureWorks LLC started with the joint bench of Cargill and Dow Chemicals in late 1980s. The capacity of their initial production of Ingeo® was 140 000 Mt per year in 2002 and renovated their facility to achieve a 25–30 % increase in annual production over the past several years. Ingeo® has various commercial grades covering the crystalline and amorphous polymers for which mainly biodegradable application has been intended. Currently (in 2012), over 20 grades are available for a variety of applications. Table 5 shows the properties of typical Ingeo® polymers.

Since NatureWorks became aware that the potential market of PLLA ought to be larger than they had considered before, they have started accelerating their business in global scale. In addition, the market needs have been directed to high-quality PLLA that can satisfy high-end applications such as apparel fibers, bottles, films, packagings, home textiles, and other durable products. This tendency has pushed NatureWorks to produce a new grade PLLA having a higher L-content (high-L PLLA) and showing a higher performance. One of the striking drives in their PLLA business is that PTT Chemical Public Co. Ltd. (PTT Chemicals) in Thailand decided to invest US$ 150 million in NatureWorks. PTT Chemicals is known as a foundation of the joint venture with Mitsubishi Chemical Corp. (Japan) to promote the production of bio-succinic acid and its polymer PBS. Their investment in NatureWorks, reflecting the strong global green concern, stimulates the growth of the green polymeric industry in Asia. Not only PLA products but also lactide monomers, which have not yet been in sales, are now lined up in the products commercialized by NatureWorks LLC. Based on these activities, NatureWorks is planning to increase the annual production of lactide and is encouraging its utilization by intentionally allowing the L-lactide users to take an advantage in using their Ingeo® packaging license.

1. 8. 2 PURAC

PURAC is a CSM's subsidiary in the Netherlands, rolling out their high quality PLLA 'Purasorb®' for medical and pharmaceutical industries. PURAC has been covering, however, only a part of the world PLA-related business before. Recently, PURAC has declared that they themselves focus on the production and sales of lactide monomers 'Puralact®' and support the production of PLA by the polymer industries in order to accelerate the formation of business network and foundation of joint works. They have already built a lactide plant of 75 000 Mt in annual production scale

Table 5 Properties of the representative commercial grade Ingeo® PLLA

Ingeo grade	MFR (g/10min, 210°C/2.16kg)	Tm (°C)	Tg (°C)
For extrusion, injection			
2003D	6	145–160	55–60
3001D	22	155–170	55–60
3251D	80	155–170	55–60
3801X	—	155–170	45
For film, sheet			
4032D	7	155–170	55–60
4060D	10	—	55–60
For fiber, non-woven			
6060D	8	125–135	55–60
6252D	80	155–170	55–60
6752D	14	145–160	55–60

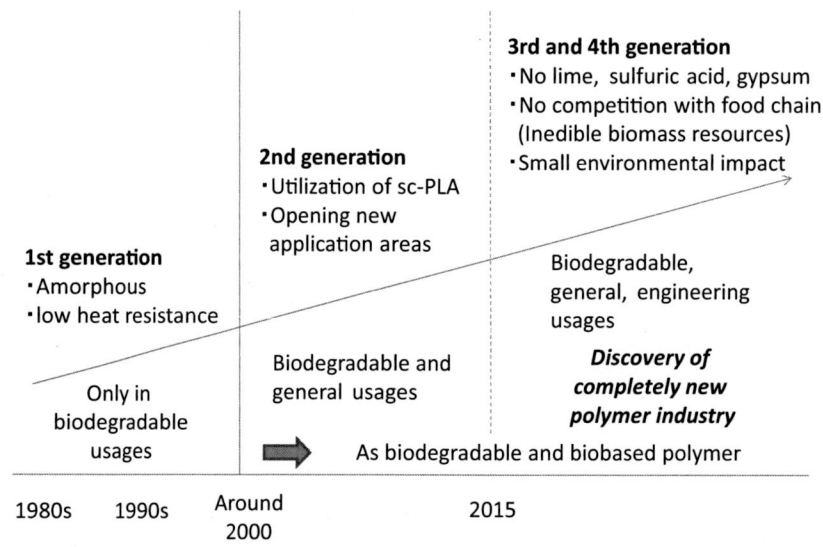

Figure 4 Development track of PLA proposed by PURAC

in Thailand in 2011 and intend to increase it worldwide. Figure 4 shows the track of future development of PLA reported by PURAC.

According to PURAC, PLA of the 1st generation has been utilized only in the biomedical application and therefore recognized as a high-cost polymer. However, the technological advance and the increasing social needs for biodegradable plastics have made it forward to its 2nd generation stage where low profitable general purpose usages have been targeted as described above. In the 3rd generation, having started around 2000, the biobased nature of PLA is focused, and therefore PLA are applied to structural materials field as high-performance biobased polymers. In the 4th generation, starting in 2015, the production of PLA will be completely independent of the edible biomass chain. In fact, PURAC improves the physical and mechanical properties of PLA polymers by using high-L PLLA and sc-PLA. They are also utilizing cellulose as the main raw feedstock in order to make their PLA production independent from the food chain, as depicted in Figure 4.

1. 8. 3 PURAC-Sulzer Chemtech-Synbra

This triangular activity is one of the several joint industrializations of PLA principally directed by PURAC. PURAC and Sulzer Chemtech (Switzerland) started their joint development of technology for reducing the polymerization cost of the PLA with better performance. They built their first polymerization plant of PLA in cooperation with Synbra in the Netherlands. This new plant has a capacity of 5,000 Mt in the annual production, and the PLA polymers produced in this plant are soley sold by Synbra. PURAC and Sulzer Chemtech intend to expand its production scale to 50 000 Mt in the nearest future. Synbra's PLA polymers are trademarked 'Synterra®', involving both enantiomeric PLLA and PDLA grades. It is expected that the commercialization of PDLA will promote the industrialization of sc-PLA showing high thermal resistivity. Moreover, Synbra has started 'Bio-foam Project' in which a new platform is intended to be established for molding, processing, and foaming PLA polymers including sc-PLA.

1. 8. 4 PURAC-Arkema

The purpose of the collaboration of PURAC and Arkema is to develop high profitable PLA materials based on block copolymer techniques, especially those with improved thermal and physical durability. In this collaboration, PURAC provides its high quality L- and D-lactides whilst Arkema provides their technologies and know-how for the precise control of polymerization. New biobased building blocks are also developed by their collaboration.

1. 8. 5 PURAC-Indorama

PURAC tied up with Indorama Ventures LLC to start a PLA business in Thailand. They aim to annually produce 100 000 Mt of PLA by developing its specific application grades that can contribute to a development of biobased textiles and packages.

In summary, Table 6 shows the market sizes estimated for three types of PLA polymers in which "PLA" denotes the amorphous or less crystalline PLLA polymers containing a considerable amount of D-lactate units. These PLA polymers are mainly used as biodegradable plastics. On the other hand, "PLLA" and "sc-PLA" are those of high crystallinity that are used as structural materials as well as fibers and textiles [21]. The market increase ought to be very high in every category, if the current development status is maintained. In particular, the use of sc-PLA may be more strongly supported in the future because of its higher thermal resistivity [22].

Table 6 Estimated market potential for PLAs suggested by McKinsey & Company (in kMt) [22]

Year	2015	2020
PLA	550	3,000
PLLA	190	1,900
sc-PLA	360	1,100

1. 9 Industrialization of sc-PLA

As described above, the sc-PLA polymers will occupy approximately 37 % of the whole market of PLA materials in 2020. Its thermal and mechanical durability have been proven by many scientific reports. However, many technical barriers are still remaining in the industrial processing of sc-PLA, because homo-chiral crystallization is likely to accompany the formation of sc-PLA when an enantiomeric mixture of PLLA and PDLA with high molecular weight is subjected to the conventional extrusion and injection moldings. To overcome this difficulty, a number of trials have been done in industry [21].

1.9.1 FKuR/Synbra

FkuR (Germany) and Synbra collaborate to develop sc-PLA to enhance the properties of PLA compounds. Their sc-PLA products are made by blending highly optically pure PLLA and PDLA to allow the high temperature application of the sc-PLA. In their official announcement, their sc-PLA shows a melting temperature (T_m) of 220 °C, which is 50 °C higher than that of the single PLLA polymer although the detail has not been unveiled. This production of sc-PLA well fits the concept of 3rd generation PLA proposed by PURAC.

1.9.2 Teijin

The brand name of Teijin's sc-PLA is 'Biofront®', which has recorded the first mass-produced sc-PLA in the biobased polymer industry. Biofront® shows excellent physical properties. For example, the T_m and heat distortion temperature (HDT) under a pressure of 0.45 MPa are 215 °C and 130 °C, respectively, while the modulus is 1,150 MPa at 23 °C. These values reach those of the petroleum-based engineering plastics such as poly(butylene terephthalate). Mazda, a Japanese motor company, once examined the Biofront® fibers for making car sheets of their hydrogen battery car in 2009. Teijin has accumulated a volume of know-how both for extrusion and injection moldings of Biofront® plastics for which PLA-containing block copolymers are used together with some additives to control the crystal structure.

1.10 Manufacturing of bacterial polyesters

PHAs are a series of aliphatic polyesters synthesized by fermentation using such bacteria as *Ralstonia eutropha* [23–25]. PHAs are crystalline polymers whose T_m is in the range of 130–175 °C. The basic polymer structure of PHAs is poly(3(R)- hydroxybutyrate) (PHB) that is the primary energy storage material of bacteria. The drawback of PHB is that it is likely to decompose around its T_m (175 °C), making its melt processing very difficult. Therefore, poly(3-hydroxybutyrate-co-3-hydroxyvalerate) (PHBV), a random copolymer with the second unit (3HV), was developed. The British company ICI launched its manufacturing in 1981 by the trade name of BIOPOL®, but was obliged to sell it after about two decades. The technology was finally transferred to Metabolix (USA), which has been trying to continue its business (Mirel® and Mvera®) with several supporting companies. Now, Metabolix seems to be interested in producing poly(3-hydroxybutyrate-co-4-hydroxybutyrate) (P3HB4HB). Even after passing 30 years from the industrialization of PHBV, the market of PHAs has remained small because of their poorer processability and properties in comparison with the ordinary plastic materials. The other noteworthy activity in manufacturing a PHA copolymer is by Kaneka (Japan). They are planning to industrialize poly(3(R)-hydroxybutyrate-co-3-hydroxyhexanoate) (PHBH) with which the property control is much easier. It was found out that by feeding *Aeromonas caviae* with a fatty acid of vegetable oil origin as a carbon source the copolymer consisting of 3HH units is produced. Kaneka developed a transgenic method to make *E. coli* produce PHBH in a scale of 1,000 Mt per year and has now started sampling work. They will increase its production in the nearest future. For larger application of these PHAs, cost-property performance ought to be improved.

1. 11 Biodegradable poly(butylene succinate) and its copolymers

Among the two-component aliphatic polyesters, poly(butylene succinate) (PBS) consisting of 1,4-butanediol and succinic acid has been known to exhibit a relatively high T_m as well as excellent physical properties. Usually, a PBS oligomer with a molecular weight of about 10 000–20 000 is first prepared by the ordinary polycondensation and then chain-extended with an isocyanate to increase the degree of polymerization [26–27]. Showa Denko (Japan) is commercializing this type of PBS by the name of Bionolle®. In addition to this PBS, chain-extended poly(butylene succinate-co-adipate) (PBSA) is also commercialized for supplying softer copolymer grades. Similarly, Mitsubishi Chemical Co. (Japan) is manufacturing poly(butylene succinate-co-L-lactate) (PBSL) containing a small amount of L-lactate units (2–3 unit mol%). Its trade name is GS Pla®. In this copolymer the molecular weight becomes higher than that of the PBS alone to exclude the chain extension process.

On the other hand, succinate or even adipate units are introduced into terephthalate polyesters to synthesize copolyesters such as poly(butylene succinate/terephthalate) (PBST), poly(ethylene succinate/terephthalate) (PEAT). These copolymers are commercialized as Ecobio® and Ecoflex® (BASF) and Apex® (Du Pont), respectively, being used as biodegradable plastics. Since these PBS derivatives exhibit a poorer performance in physical properties, they are utilized for making mulching films and garbage bags, all of which are for biodegradable purposes. At present, both the succinic acid and 1,4-butanediol are oil-based in their origin, and any of the PBS polymers and copolymers manufactured are not biobased. However, the production of succinic acid will be replaced by the biobased route as described above, and even 1,4-butanediol will possibly be produced biologically. Therefore, PBS will be fully biobased polymers in the near future.

1. 12 Other biobased polymers

1. 12. 1 Furanics and biobased terephthalic acid

Furan-based building blocks such as 2,5-furandicarboxylic acid (FDCA), hydroxymethylfurfural, and furfural are called furanics. YXY® (pronunciation: icksy) is the brand name for the furanics that are produced by Avantium (The Netherlands) for use as biofuels, green chemicals, and green materials [28]. The potentially enormous marketability of furanics is evidenced by Du Pont and DSM. In addition, furanics are lined up as the biobased platform chemicals proposed by DOE (USA). Avantium has started collaboration with NatureWorks, Teijin Aramid, Solvay, and Coca-Cola to develop new biobased plastics for green and recyclable applications. The most important is FDCA which is a building block of poly(ethylene furanoate) (PEF) [29]. PEF can be an alternative of poly(ethylene terephthalate) (PET) to make high profitable biobased plastic bottles, because it is characterized by its high level of oxygen barrier although the other properties are not as good as those of PET in general. The collaboration with Coca-Cola in developing YXY is, therefore, most striking to cast a possibility of the current petroleum-based bottle industry spinning out. The first industrial-scale YXY plant will be in operation in 2015 in a capacity of 30 000–50 000

Mt in annual production.

Coca-Cola intends to accelerate their green bottle production. Since 2009, they have already introduced biobased ethylene glycol for making the bottle PET, and by which the biobased content of the bottle amounts to 30 wt% now. This activity by Coca-Cola may afford a key for smoothly replacing the petroleum-derived plastics by biobased plastics in the bottle industry. Their strong partners other than Avantium are Virent (USA) and Gevo (USA) who are engaged in developing biobased *p*-xylene and terephthalic acid from carbohydrate feedstock. Virent has a new technology to convert biomass-derived sugar into *p*-xylene as well as to other types of chemicals and biofuels whose commercial name is BioFormPX®. The investors to Virent involve Shell, Cargill, and Honda (Japan) in addition to Coca-Cola. Gevo has a fermentation technology based on their original yeast that can efficiently convert sugars into isobutanol. They first convert this isobutanol into isobutylene by dehydration, and in the following step the isobutylene is dimerized into isooctene, which is finally subjected to dehydrocyclization reaction to make alkylbenzene containing approximately 90 % of *p*-xylene (Figure 5(b)). The *p*-xylene thus obtained can readily be converted to terephthalic acid that is used for making fully biobased PET when combined with bio-ethylene glycol. The fully biobased PET bottles are cost-competitive with the conventional PET bottles, affording a "green value" to the current bottle industry. Gevo is expanding their isobutanol plant to produce 38 million gallon of isobutanol in 2013, which may be enough for producing bio-PET. Another strong joint venture for the production of isobutanol is Butamax™ Advanced Biofuels LLC (USA) sponsored by Du Pont and BP. The main biobased product of Butamax is 1-butanol for use as biofuel. This 1-butanol can readily be converted into isobutanol corresponding to the market needs.

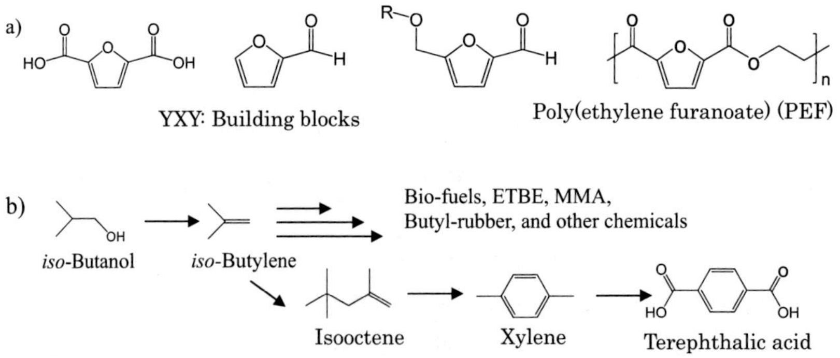

Figure 5 (a) Chemical structures of YXY® and PEF synthesized by Avantium and (b) synthetic routes to biobased terephthalic acid proposed by GEVO

1. 12. 2 Various biobased polyamides

Very recently, active research and development have been conducted for biobased polyamides by using a wide variety of diamines, dicarboxylic acids, and lactams that can be obtained from biobased feedstock. Succinic acid and amino acids that are readily synthesized by fermentation can be utilized as the starting substances. Various fully and partly biobased polyamides having different properties can be in hand when these monomers are combined with each other and with the conventional non-biobased ones [29–30]. Since the costs of the ordinary polyamides such as Nylon 66 and Nylon 6 are rather expensive compared with those of polyesters, it should be easier to compensate the monomer costs that are likely to become higher when they are prepared by the biobased routes.

Table 7 summarizes the polyamides currently developed. Some of these polyamides are fully biobased, while the others are only partly biobased because either the diamine or dicarboxylic acid combined is oil-based.

The biobased monomers such as γ-aminobutyric acid (C4), succinic acid (C4), 1,4-butanediamine (C4), 1,5-pentanediamine (C5) can be obtained by microbial fermentation with sugar feeds, whereas sebacic acid (C10) and 11-aminoundecanoic acid (C11) are synthesized from castor oil via ricinoleic acid (C18) [31]. 1,10-Decanediamine is also synthesized from sebacic acid by the reduction of sebacadiamide. Among these monomers, sebacic acid is combined with various diamines to prepare fully and partly biobased polyamides such as nylon 410, nylon 610, nylon 510, nylon 1010, and poly(xylene sebacamide). On the other hand, nylon 11 is synthesized by polycondensation of 11-aminoundecanoic acid, whereas nylon 4 is synthesized by the ring-opening

Table 7 Fully and partly biobased polyamides being developed

Nylon 4	γ-Aminobutyric acid	AIST
Nylon 46	1,4-Diamionobutane	DSM (*Stanyl*)
	Adipic acid	
Nylon 410	1,4-Diamionobutane	DSM (*EcoPaXX*)
	Sebacic acid	
Nylon 56	1,5-Pentanediamine	Toray
	Adipic acid	
Nylon 510	1,5-Pentanediamine	Toray
	Sebacic acid	
Nylon 64	1,6-Hexanediamine	BASF
	Succinic acid	
Nylon 610	1,6-Hexanediamine	Toray, BASF, Du Pont, Daicel-Evonik (*VESTAMIDR Terra HS*)
	Sebacic acid	
Nylon 1010	1,10-Decanediamine	Du Pont, Daicel-Evonik (*VESTAMIDR Terra DS*)
	Sebacic acid	
Nylon 11	11-Aminoundecanoic acid	Arkema (*Rilsan*)
Nylon 4&6	γ-Aminobutyric acid	KRICT
	ε-Caprolactam	
Poly(xylenesebasamide)	Xylenediamine	Mitsubishi Gas Chemical (*Lexter*)
	Sebacic acid	
Nylon 4T, Nylon 10T, Nylon XT	α,ω-Alkylenediamine	Unitika (*Zecot*), Toyobo (*Vyloamide*), Daicel, DSM
	Terephthalic acid	
Aramide	2-Pyrone-4,6-dicarboxylic acid	(proposed)
	m-Phenylenediamine	

polymerization of 2-pyrrolidone which is a cyclic product of γ-aminobutyric acid obtained from glutamic acid. These polyamides have rather specific applications as soft and hard Nylons for electronic and car parts.

1. 13 Latest examples of high-performance biobased polymers

A considerable number of biobased polymers have already been synthesized as mentioned above. Their practical application, however, has been limited to biodegradable, general, or very-low-grade engineering usages yet. Much effort has therefore been made to develop high-performance and specialty biobased polymers, i.e., for highly profitable engineering usages. The following examples are those for the high specialties and functionalities.

1. 13. 1 Bio-based LCP

Biobased liquid crystalline polymers (Bio-LCP) consisting of aromatic building blocks of natural origin were synthesized as high-performance polymers. Typically, 4-hydroxycinnamic acid (4HCA), which is an intermediate metabolite in the biosynthetic pathway of lignin in plant cells, is polymerized into a Bio-LCP [32]. The mechanical properties of the 4HCA-derived Bio-LCP are highly superior to those of the currently commercialized biobased polymers in terms of mechanical strength (σ = 63 MPa), Young's modulus (E = 16 GPa), and maximum softening temperature (169 °C). In addition, their mechanical properties can be improved by light irradiation.

1. 13. 2 Isosorbide-containing polymers

Isosorbide is a bicyclic compound which is produced from sorbitol, one of the platform chemicals. In general, glucose is hydrogenated into sorbitol having six hydroxyls. Its cationic dehydration undergoes intra-molecular ether formation to give isosorbide. Isosorbide is commercialized by Roquette (France), UENO Chemicals (Japan), and other companies. Since isosorbide has two hydroxyls, it can be a building block of polyesters and polycarbonates [33]. The most interesting is the isosorbide-based polycarbonate that is commercialized by Mitsubishi Chemical Co. (Japan) by the trade name of "Durabio®". It retains similar properties with the conventional bisphenol A-type polycarbonate (Tg ~ 150°C), whilst showing better optical properties. Durabio® is convinced to have great potential for replacing the conventional polycarbonate.

Duchateau et al. reported that isosorbide can be incorporated into poly(butylene terephthalate) by solid-state polycondensation (SSP) of a prepolymer composed of isosorbide, terephthalic acid, and 1,4-butanediol [34]. As a result of milder polymerization condition in the SSP, thermal degradation and trans-esterification could be avoided, and a high-molecular-weight polymer having a number average molecular weight up to 100 000 g/mol was produced with minimum coloration of the polymer. The SSP products exhibited a blocky structure and showed better thermal properties than the conventional melt polymerization products: the T_g, T_m, and T_c were significantly higher in the former. Koning et al. reported that SSP can be a powerful synthetic method to obtain a high-molecular-weight polyamide consisting of sebasic acid, 1,4-butanediamine, and a diaminated isosorbide [35]. In their experiment, short time polycondensation followed by the SSP was shown to produce a white polymer whose number average molecular weight was above 18 000 g/mol. Its T_m was successfully controlled between 150 and 240 °C by changing composition of the isosorbide unit. Isosorbide can also be incorporated into PET to increase the T_g value (Figure 6). When the isosorbide content is increased by 1 unit mol% in a range from 1 to 10 unit mol%, the T_g value of the isosorbide-containing PET increases by 1 °C. The increased thermal durability of the polyesters enables us to develop "hot fill" bottle which is not obtainable with the conventional PET [36, 37].

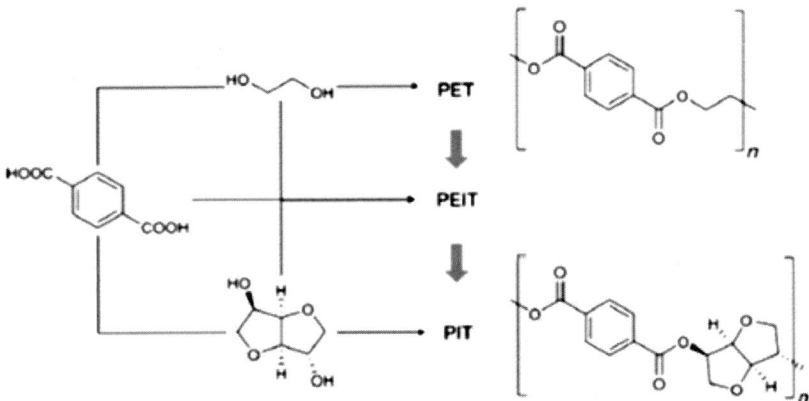

Figure 6   Incorporation of isosorbide into PET

1. 13. 3 Terpene-derived polymers

Terpenes are C10 homologues primarily produced by plants as conifers. Since terpenes are composed of isoprene units (C_5H_6), their polymerization may not give crystalline polymers with high performance. It is however possible to obtain high T_g polymers by using naturally occurring terpenes as building blocks. It has recently been demonstrated that modified terpenes can be polymerized into various bio-based polymers and copolymers. For example, Kamigaito et al. conducted cationic polymerization of (2)-β-pinene and (2)-α-phellandrene to produce alicyclic polyolefins with high T_g (>100 °C) and excellent transparency (Figure 7(a)) [38]. Myrcene, easily produced by pyrolysis of pinene, was also shown to be another candidate monomer for making biobased polyolefins (Figure 7 (b)) [39]. For example, a cyclic diene monomer, 3-methylenecyclopentene, was prepared by ring-closing metathesis of myrcene. It was polymerized under radical, anionic, and cationic

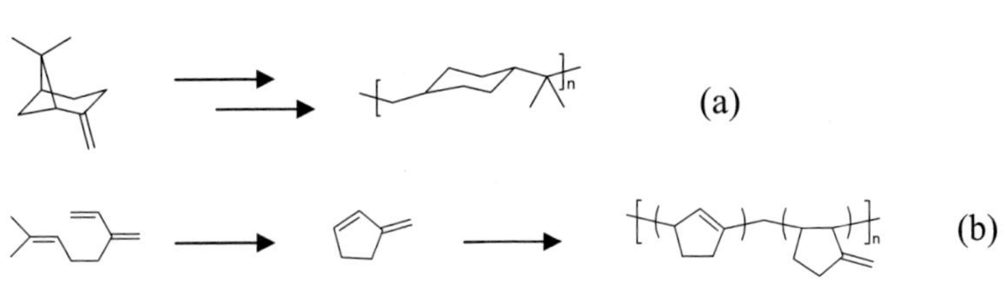

Figure 7   Polymerizations of (a) β-pinene and (b) 3-methylenecyclopentene synthesized by metathesis of myrcene.

conditions to give polymers showing T_g and T_m at -4 – 11 °C and 65 – 105 °C, respectively.

Myrcene was also polymerized with a special cationic catalyst to give a myrcene polymer having controlled stereo structure, molecular weight, and polydispersity.

1. 13. 4 Specialty PLA polymers from modified lactides

Since the general PLA polymers show average properties, they have been modified for imparting specialties with which their application to engineering and biomedical fields may be possible. The first approach is to substitute the methyl group of lactide monomers by other functional groups. One of the most successful is mandelide having phenyl groups instead of methyl groups. Mandelide is readily prepared by dehydration of mandelic acid that can be extracted from bitter almonds and is known as a useful precursor of various pharmaceuticals. It can be polymerized by ROP as lactide monomers to produce polymandelide (PMA), which is a glassy polymer showing a high T_g (~100 °C) with transparent nature (Figure 8 (a)) [40]. The properties of PMA are similar to those of polystyrene. Another intriguing trial to modify lactides for high-Tg polymer was reported by Hillmyer et al [41]. They first brominated L-lactide and then conducted hydro-bromide elimination to synthesize (6S)-3-methylene-6-methyl-1,4-dioxane-2,5-dione. This modified compound was then reacted and a pentadiene to form a tricyclic compound by the Diels-Alder coupling. The ring-opening metathesis polymerization of this tricyclic monomer was finally conducted to obtain a high-molecular-weight polymer with narrow polydispersity (Figure 8 (b)). The T_g of the resultant polymer reached 192 °C, indicating a high-performance nature. Another interesting example was related with a water-soluble PLA polymer which was reported by Baker et al. In their approach, the methyl groups of lactide were substituted by water soluble poly(ethylene glycol) (PEG) chains (Figure 8 (c)) [42]. The obtained PEG-modified lactide was subjected to the ordinary ROP to obtain a polymer having high hydrophilicity. Interestingly, the polymer exhibited a lower critical solution temperature (LCST) near ambient temperature. This unique LCST behavior suggests its potential application to the biomedical and pharmaceutical fields.

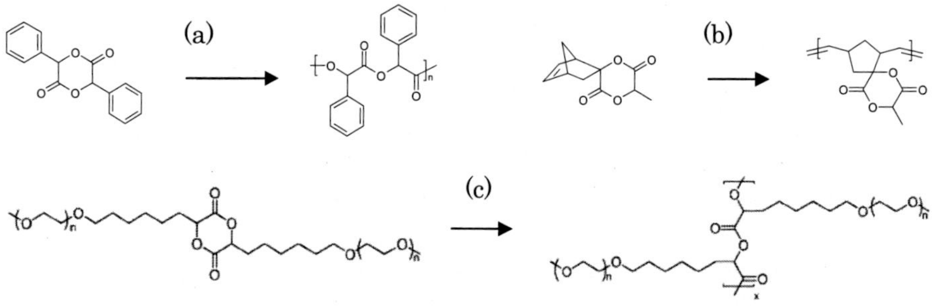

Figure 8 Syntheses of modified lactide polymers from (a) mandelide, (b) (6S)-3-methylene-6-methyl-1,4-dioxane-2,5-dione, and (c) poly(oxyethylene)-substituted lactide

1. 14 New platform of biobased polymers

The above examples strongly suggest that a novel polymer platform may be established in the near future if many other biobased polymers are developed from different biomass resources. Figure 9 shows the classification of the current bio-based polymers as compared with that of the conventional oil-based polymers. It is evident that most of the bio-based polymers are located in the lower two categories in regard to performance, i.e., general purpose and temporal use. Very few high-performance biobased polymers that can belong to the upper two categories have been developed, probably because of the deficiency of aromatic building blocks. While the value-chains of the current bio-based polymers are being established, much effort must be made to create high-performance polymers that are suited for application to high value-added area. With this effort we will finally establish a new platform of polymeric materials that will lead innovation of the polymer science and technology in this century.

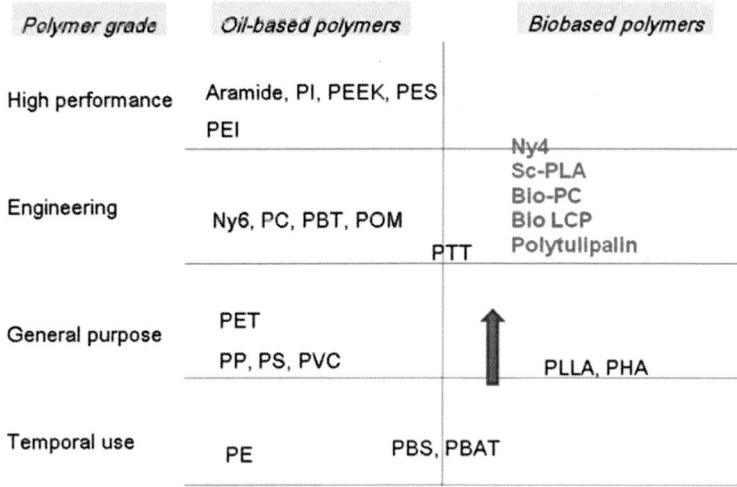

Figure 9   Creating a new platform of high-performance biobased polymers corresponding to that of oil-based ones.

References

1. S. S. Im, Y. H. Kim, J. S. Yoon, I.-J. Chin, Eds., 'Bio-Based Polymers: Recent Progress', Wiley-VCH, Weinheim, Germany (2005)
2. Y. Kimura, *Polym. J.*, **41**, 797–807 (2009)
3. M. Vert, S. M. Li, G. Spenlehauer, P. Guerin, *J. Mater. Sci. Mater. Med.*, **3**, 432–446 (1992)
4. M. Steib, B. Schink, *Arch. Microbiol*, **154**, 253 (1984)
5. A. Steinbuechel, 'Biopolymers', Wiley-VCH, Weinheim (2001)
6. R. A. Gross, C. Demello, R. W. Lenz, H. Brandl, R. C. Fuller, *Macromolecules*, **22**, 1106–1115 (1989)
7. J. V. Kurian, *J. Polym. Environm.*, **13**, 159–167 (2005)

8. S. Atsumi, T. Y. Wu, E. M. Eckl, S. D. Hawkins, T. Buelter, J. C. Liao, *Appl. Macrobiol. Biotechnol.*, **85**, 651–657 (2010)
9. R. Mullin, *Chem. Eng. News*, Nov., **8**, 29–37 (2004)
10. L. Shen, J. Haufe, M. K. Patel, 'Product overview and market projection of emerging bio-based plastics PRO-BIP 2009', Utrecht University, The Netherlands (2009)
11. T. Werpy and G. Petersen, "Top Value Added Chemicals from Biomass", The National Renewable Energy Laboratory & DOE National Laboratory (2004)
12. H. Mutlu1, M. A. R. Meier, *Eur. J. Lipid Sci. Technol.*, **112**, 10–30 (2010)
13. R. Kirschbaum, personal communication in 2012.
14. J. V. Kurian, 'Natural Fibers, Biopolymers, and Biocomposites', **15**, 487–525, CRC press (2005)
15. H. Yano, Sen'i Gakkaishi, **62**, 356–358 (2006)
16. S. K. Ritter, *Chem. Eng. News,*, Feb., **14**, 11–15 (2011)
17. http://www.dsm.com/en_US/cworld/public/sustainability/downloads/publications/oct_2010_dsm_roquette_bio_based_sustainable_succinic_acid.pdf
18. I. T. Tong, H. H. Liao, D C Cameron, *Appl Environ Microbiol.*, **57**(12), 3541–3546 (1991)
19. http://www.crops2industry.eu/images/pdf/bordeaux/9.%20RHODIA.pdf
20. R. E. Drumright, P. R. Gruber, D. E. Henton, *Adv. Mater.*, **12**, 1841–1846 (2000)
21. M. Kakuta, M. Hirata, Y. Kimura, *J. Macromol. Sci., Part C: Polym. Rev.*, **49**, 107–140 (2009)
22. http://www.csmglobal.com/_sana_/handlers/getfile.ashx/e577bd67-b9c2-4002-bc65-d0f9906dc5b4/Purac+seminar+31-3-2010+-+Biopolymers+Business+opportunities+for+a+sustainable+future.pdf
23. Y. Doi, A. Tamaki,; M. Kunioka, K. Soga, *Appl. Macrobiol. Biotechnol.*, **28**, 330–334 (1988)
24. Y. Saito, Y. Doi, *Int. J. Biol. Macromol.*, **16**, 99–104 (1994)
25. Y. Doi, S. Kitamura, H. Abe, *Macromolecules*, **28**, 4822–4828 (1995)
26. I. Taniguchi, S. Nakano, T. Nakamura, A. El-Salmawy, M. Miyamoto, Y. Kimura, *Macromol. Biosci.*, **2**, 447–455 (2002)
27. M. Inui, S. Murakami, S. Okino, H. Kawaguchi, A.A. Vertes and H. Yukawa, *J. Mol. Microbiol. Biotechnol.*, **7**, 182–196 (2004)
28. http://nanofactory.org.uk/admin/uploads/file/111006%20Shefield_Avantium.pdf
29. A. Gandini, D. J. A. Silvestre, C. N. Pascoal, F. A. Sousa, J. M. Gomes, J. Polym. Sci. Part A, *Polym. Chem.*, **47**, 295–298 (2009)
30. S. Kind and C. Wittmann, *Appl. Microbiol. Biotechnol.*, **91**, 1287–1296 (2011)
31. Sebacic acid
32. T. Kaneko, M. Matsusaki, T. T. Hang, M. Akashi, *Macromol. Rapid Commun*, **25**, 673–677 (2004)
33. Noordover, B. A. J.; Haveman, D.; Duchateau, R.; Benthem, R. A. T. M.; Koning, C. E. J. *Appl. Polym. Sci.*, **121**, 1450–1463 (2011)
34. R. I. Sablong, R. Duchateau, C. E. Koning, G. Wit, D. V. Es, R. Koelewijn, J. V. Haveren, *Biomacromolecules*, **9**, 3090–3097 (2008)
35. L. Jasinska, M. Villani, J. Wu, D. V. Es, E. Klop, S. Rastogi, C. E. Koning, *Macromolecules*, **44**, 3456–3466 (2011)
36. M. Rose, R. Palkovits, *ChemSusChem.*, **5**, 167–176 (2012)
37. J. C. Bersot, N. J. R. S-Loup, P. Fuertes, A. Rousseau, J. P. Pascault, R. Spitz, F. Fenouillot, V. Monteil, *Macromol. Chem. Phys.*, **212**, 2114–2120 (2011)
38. K. Satoh, H. Sugiyama, M. Kamigaito, *Green Chem.*, **8**, 878–882 (2006)
39. S. Kobayashi, C. Lu, T. R. Hoye, M. A. Hillmyer, *J. Am. Chem. Soc.*, **131**, 7960 (2009)
40. T. Liu, T. L. Simmons, D. A. Bohnsack, M. E. Mackay, M. R. Smith, G. L. Baker, *Macromolecules.*, **40**, 6040–6047 (2007)
41. F. Jing, M. A. Hillmyer, *J. Am. Chem, Soc.*, **130**, 13826–13827 (2008)

42. X. Jiang, M. R. Smith III, G. L. Baker, *Macromolecules.*, **41**, 318–324 (2008)

Other information sources

Bioplastic Magazine., http://www.bioplasticsmagazine.com/en/index.php
Chemistry and Engineering News., http://cen.acs.org/index.html
European Bioplastics., http://www.europeanbioplastics.org/
Japan Bioplastics Association., http://www.jbpaweb.net/
US Department of Energy., http://energy.gov/

Chapter 2
Biomass and Biomass Refining

Maria José Climent, Maria Mifsud and Sara Iborra

2.1 Introduction

Currently, the three quarters of the word's energy supply comes from fossil feedstocks, being particularly petroleum an essential building block of our economy. By refining crude oil, fuels and a wide variety of chemical compounds are produced. However, the growing scarcity of these limited resources, the increase in oil prices, concerns regarding the dramatic increase of atmospheric CO_2 levels, and the projected increasing demand for energy in the future, are driving society towards the search for new sustainable sources of energy and chemicals. In recent years, biomass, the only renewable source of organic carbon, has been pointed out as the perfect equivalent to petroleum for the production of fuels, chemicals and carbon-based materials. However, while oxygenated nature and the chemical disparity of biomass appears as the main drawback for the production of biofuels, their highly functionalised nature gives biomass high potential for the production of chemicals. Thus, the production of chemicals from fossil feedstocks involves a previous selective functionalization step in order to activate the initial feedstock for subsequent upgrading processes. However, the highly functionalized molecules coming from biomass show a rich chemistry which offers high flexibility to be converted into a large variety of useful compounds. Considering that only 5% of the total oil production is devoted to the production of chemicals, it has been estimated that the amount of non-edible biomass available (lignocellulosic biomass) could theoretically replace fossil feedstocks for this purpose [1].

The success of the petroleum industry is based on the complete utilization of petroleum resources by optimized technologies in highly integrated facilities. Thus, the economic feasibility of oil refineries is achieved by the simultaneous production of large volumes of low-value transportation fuels along with lower volumes of more costly chemicals. In fact, most of chemicals and specialty products that enhance the quality of our everyday life, are derived from a few molecules (benzene, toluene, xylene, ethylene, propylene, butadiene and methanol) which serve as building blocks for the petrochemical industry. A similar approach could be followed by a future biomass-based industry, where a variety of bio-feedstocks could be converted into biofuels, valuable chemicals and energy, through the processing of selected biomass derivatives (the so-called platform molecules) in an integrated facility, called biorefinery.

In 2004, the US Department of Energy (DOE) published the first of two reports outlining research needs for biobased products. The report described a group of 15 target compounds (known as the DOE "Top 10" report) that could be produced from biorefinery carbohydrates [2]. Organic acids such as succinic, itaconic, fumaric, lactic and levulinic, and polyols such as xylitol, sorbitol and glycerol were some examples of platform molecules presented in the original list. However, six years after, considerable progress in biobased products has been made, and recently Bozell and Petersen [3] revisited this list and proposed an updated list of compounds which included many of those originally selected as well as others (e.g. ethanol, bio-hydrocarbons, and furans). In last years, a number of catalyzed processes (dehydrations, oxidations, hydrogenations, hydrogenolysis, isomerizations, etc) have been proposed to produce a large number of molecules starting from these biomass-derived building blocks [4]. In this chapter we will describe the chemical or biochemical production of some of these platform molecules (carboxylic acids, furans, and polyols) and its conversion of into useful chemicals, putting special attention on its transformation into biobased polymeric materials.

2.2 Carboxylic acids and polyols

2.2.1 Lactic acid

Lactic acid (LA, 2-hydroxypropanoic acid) is the most extensively existing carboxylic acid in the nature [5]. The word's annual production is around $3.5 \: 10^5$ tons/year and its grow is believed by some observers to be 12–15% per year.

LA can be produced by chemical synthesis or by fermentation of different carbohydrates such as glucose, maltose, sucrose, lactose, etc. However, lactic acid is commercially produced nowadays mainly though the fermentation of glucose. One of the major drawbacks in the lactic acid production *via* fermentation process is their recovery from the broth. The conventional process involves the precipitation of calcium lactate after the separation of the microorganisms and the conversion of the salt to lactic acid by addition of sulfuric acid. The separation and purification stages account for up to 50% of the production cost. However, new technologies that have recently emerged can overcome major barriers in separations and purification steps. Advances in membrane based separation and purification technologies particularly in microfiltration, ultrafiltration and electrodialisis have given very promising results showing a strong potential for an efficient and economic process for recovery and purification of lactic acid without generating a salt waste [6].

Recently has been reported that acid lactic can be also synthesized directly from sugars (e.g. fructose, glucose and sucrose) using solid Sn-Beta zeolites as catalysts [7]. This can be a promising technology with clear improvements over fermentation process by facilitating separation of the lactic acid and recycling of the solid catalyst.

Traditionally, the main use of lactic acid is in food and food related applications. However since the early 1990s the lactic acid market has been expanding as a result of the development and commercialization of new applications. These new uses are related to the industrial production of polymers and chemicals [8].

The bifunctional character of lactic acid (e.g.-OH and -COOH groups) allows a variety of transformations to useful products (currently obtained from petroleum). The main derivates are: Polylactic acid, acrylic acid, propilenglycol, lactate esters, acetaldehyde and propanoic acid which are briefly discussed bellow (Scheme 1).

The primary use for lactic acid is the production of polylactic acid (PLA). Although lactic acid can undergo direct polymerization, the process is more effective if lactic acid is first converted to a low molecular weight pre-polymer (MW ~ 5000) and then depolymerized to the lactide. A wide range of catalysts based on transition metals such as tin, aluminium, lead, zinc, bismuth and yttrium are known to promote the lactide polymerization [9].

Acrylic acid, is obtained by dehydration of lactic acid. It is an important raw material for the production of plastics, adhesives and paints. Acrylic acid, its amide and ester derivates are the primary building blocks in the manufacture of acrylate polymers. Numerous applications in surface coatings, textiles, adhesives, paper treatment, leather, fibers, detergents, etc, are known. Currently 100% of acrylic acid is produced out of fossil oil, mostly via direct oxidation of propylene, therefore the production of acrylic acid via dehydration of lactic acid is an attractive target for new bio-based compounds, and most lactic acid transformations are based on this reaction [10]. Unfortunately under thermal or acid catalyzed conditions other reactions such as decarbonylation/decarboxylation processes compete with dehydration decreasing the selectivity process. Thus, different catalytic systems such as $CaSO_4/Na_2SO_4$ [11], $AlPO_4$ [12], Na_2HPO_4 supported on silica alumina [11] have been used to perform this reaction with variable success. Owing to salts and esters of lactic acid dehydrate more easily than the corresponding acid better yields can be obtained from these precursors. Thus, using aluminophosphate as a catalyst Paparizos et al [13] reported 61% yield of acylic acid starting from ammonium lactate. Walk up et al [14] developed a process for converting lactate esters to acrilate using $CaCO_4$ [14]. Zeolites were also used as catalyst to obtain acrylate [15]. Biocatalyst approach has been also developed using lactyl-CoA dehydratase but the acrylate yield in this case was very low [16].

Propilenglycol (1,2-propanediol) is a commodity chemical with a variety of uses as a solvent for the production of unsaturated polyester resins, drugs, cosmetics and foods. It is currently produced by hydration of propylene oxide, process involving the use of hypochlorous acid. However, production of propylene glycol by direct hydrogenation of lactic acid or lactates can be an alternative green route to the petroleum based processes. Adkins et al [17] described a method, using copper/chromium oxide and Raney nickel catalyst achieving 80% yield, however high temperatures and hydrogen pressure were required. Other materials based on rhenium, ruthenium and copper as free metal as well as supported catalysts have been developed in order to perform the hydrogenation reaction [8, 10, 18].

Lactate esters, from lactic acid and alcohols (particularly methanol, ethanol and butanol) are non toxic and biodegradable. They are high boiling liquids having excellent solvent properties and could potentially replace toxic and halogenated solvents for a wide range of industrial uses. Lactate esters are also used as plasticizers in cellulose and vinyl resins and enhance the detergent properties of ionic surfactant. These compounds are even useful in the preparation of herbicidal formulations [19]. For esterification of lactic acid both uncatalyzed and catalyzed processes have been described in literature. The catalytic production of lactates has been traditionally performed with conventional homogeneous catalysts, while the use of heterogeneous catalysts such as cation exchange resins [19a, 20] is comparatively scarce.

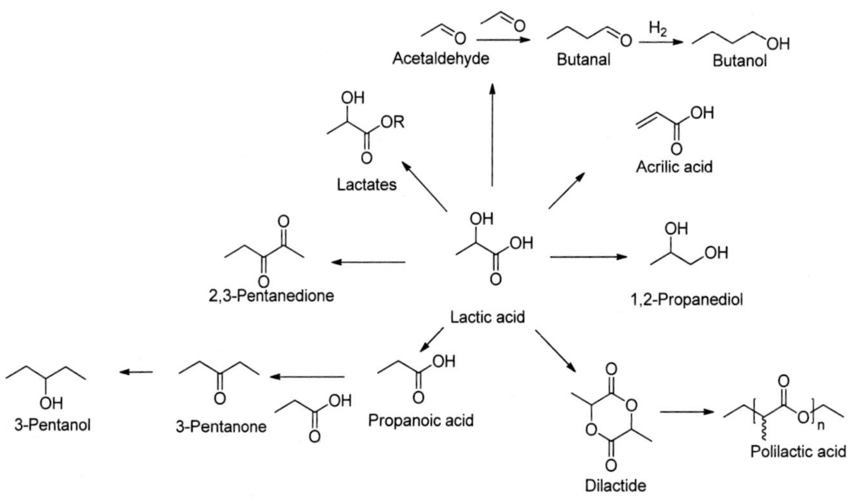

Scheme 1 Lactic acid platform

LA can be deoxygenated to obtain acetaldehyde via decarboxylation or decarbonylation processes. Acetaldehyde is widely used in the synthesis of fine chemical as well as in the bulk chemical industry. One interesting approach is the one pot synthesis of butanol in a process which involves the self-aldol condensation of acetaldehyde to generate butanal followed by a hydrogenation to obtain the corresponding alcohol using a multifunctional Pt/Nb$_2$O$_5$ catalyst [21].

Propanoic acid, which is extensively used as intermediate in organic synthesis, can be obtained as well from lactic acid by consecutive dehydration and reduction reactions. Additionally it can be catalytically converted into 3-pentanone by a C-C coupling reaction and afterwards hydrogenated to the corresponding alcohol. C$_4$-C$_5$ alcohols are suitable as high energy-density fuels for the transportation sector [21].

2,3-Pentanedione can be obtained from LA condensation. This compound is widely used as a solvent for paints, inks and lacquers and also employed as a constituent of synthetic flavoring agents. Wadley et al [22] reported the synthesis of 2,3-pentanedione from lactic acid with yields of nearly 60% of with 80% selectivity, using potassium and cesium salts supported on silica as catalysts in a fixed bed reactor and from lactic acid in the vapor phase.

2.2.2 Succinic acid

Succinic acid (SA, 1,4-butandicarboxylic acid) is among the new bio-derived building-block chemicals that could replace the current maleic anhydride C$_4$ platform. The potential market for succinic acid and its derivatives has been projected to be as much as 2.4 10^5 tons/year within next years [23].

SA can be obtained from different biomass feedstocks including wheat, corn waste and others through fermentation of glucose [24]. The optimization of the fermentation process for succinic acid production is currently underway. To that end, various natural succinate producing strains of bacteria and engineered *Escherichia coli* microorganisms have been investigated [25]. However, among the bottleneck problems of the industrial production of succinic acid from renewable resources remain the costs related to their purification [26]. Furthermore other organic acids produced as side products during the fermentation can unfavorably affect the recovery of succinic acid. In order to overcome this limitation, several strategies have been developed such as electrodialysis or direct transformation to valuable derivates *in situ* in the fermentation broth [27].

SA could be directly used for the production of some new biodegradable polymers such as nylons or polyesters [28]. For example, combination of SA and 1,4-butanediol (BDO) leads to poly-butylsuccinate (PBS), a biodegradable synthetic polyester materials with similar properties to PET which is used in biodegradable packaging films and disposable cutlery. A recent publication describes the conversion of succinic acid into new polyester for coating applications upon polymerization with isosorbide, a renewable building block also available in high yield from glucose [29].

The world market of SA derivatives (Scheme 2) includes 1,4-butanediol (BDO), tetrahydrofuran (THF), γ-butyrolactone (GBL), methyl and ethyl succinates, and also pyrrolidone and its derivatives. These compounds have important applications as solvents, additives in cosmetics and fragrances. 1, 4-Butanediol is mainly used as starting material for the production of important polymers such as polyesters, polyurethanes and polyethers [10]. Polybutylene terephthalate, is the major BDO based polymer which is mostly used for engineering plastics, fibers, films, and adhesives. Tetrahydrofuran is used as a solvent for poly (vinyl chloride) (PVC) synthesis. Moreover, is also employed as a monomer for the manufacture of polytetramethylene glycol, which is an intermediate for Spandex fibers and polyurethanes. GBL, BDO, and THF can be obtained by hydrogenation of SA using a variety of metal supported catalysts, although the typical reduction catalysts are based on copper. The selective reduction of SA to GBL, BDO and THF can be achieved by careful selection of the catalyst and reaction conditions. THF formation seems to be favoured at high temperatures and low pressures of hydrogen (2–5MPa) whereas lower temperatures and higher hydrogen pressures (4–10MPa) lead to high BDO yields [30].

Other very interesting SA derivatives are 2-pyrrolidone and N-alkyl-2pyrrolidones. The largest application of 2-pyrrolidone is the production of N-vinyl-2-pyrrolidone polymers (polyvinylpyrrolidone), with a wide range of applications in pharmaceuticals, cosmetics, textiles, paper, detergents and beverages. N-methyl-2-pyrrolidone is another important compound widely used as a solvent. 2-Pyrrolidone is usually obtained from GBL and ammonia at elevated temperatures (> 273°C) and pressures (50–160 bar), the pressure demands being lower on the use of a catalyst [31]. However, a low-cost, effective route is the direct reductive amination of SA and its derivatives to pyrrolidones in a one-pot reaction at 12–13 MPa and 235–275 °C using noble metal supported catalysts [32]. In this sense, it is possible to selectively produce pyrrolidones in high yields by using an Au-based catalyst in a one-pot process starting from succinic anhydride [33].

Succinic acid reacts with alcohols in the presence of acid catalysts to form dialkyl succinates [34]. The esters of low molecular weight (methyl and ethyl succinates) have applications as solvents and as synthetic intermediates. Many research efforts have been done to substitute the homogeneous catalyst traditionally employed for more eco-friendly heterogeneous catalyst such as Amberlist-15 [35], aluminophosphate [36], zeolites [37], sulfated zirconia and titania or Nafion-H [38]. Additionally, several research efforts are focused on the development of water stable

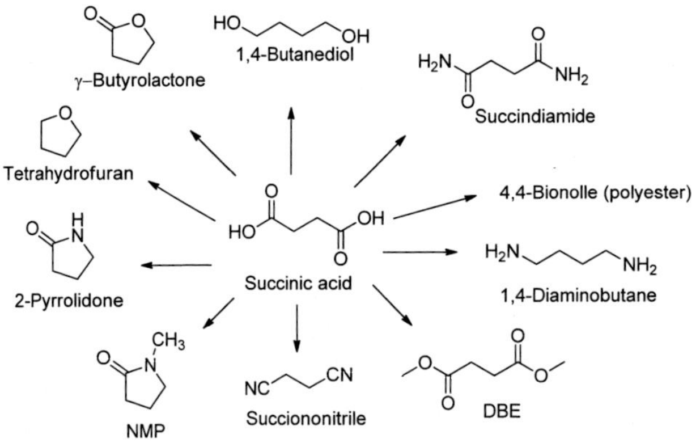

Scheme 2 Succinic acid platform

and highly active catalyst for the direct aqueous esterification. Thus, water tolerant catalysts such as Starbon® materials have been developed [39]. The water compatibility of this catalytic system allows the possibility to work with dilute aqueous mixtures in which less water soluble formed products (e.g. esters, amides, lactones) are easily separated out from the original mixture [27].

2.2.3 3-Hydroxypropionic acid

3-Hydroxypropionic acid (3HPA) is a structural isomer of lactic acid. 3HPA can be produced from a wide range of compounds through chemical synthesis, but its production is not commercially feasible due to various technological and environmental concerns [40]. This provide a clear incentive for a biological production of 3HPA, and recently the production of 3-hydroxypropinoic acid based on glucose and/or glycerol fermentation [41] has gained considerable attention.

Due to its bifunctional character, the potential applications of this platform molecule is enormous (Scheme 3), ranging from its direct use to as a precursor for the synthesis of other commercially valuable chemicals such as 1,3-propanediol, acrylic acid, acrylamide or malonic acid [42].

3HPA can be polymerized to poly-3-hydroxypropionic acid which is a bio-derived plastic with interesting physical characteristics as a high glass transition temperature and a melting point of 170–184 °C, which means unusual heat stability [43]. Additionally, the polymer is enzymatically and hydrolytically degradable, thus making it more environmentally attractive [44].

The acid group of 3HPA can be reduced giving 1,3-propanediol which is a starting material in the production of polyesters. Currently the production of this diol is based on fossil feedstock and one of its main applications is the production of the polytrimethylene terephtalate (PTT), by reaction with terephthalic acid. This polymer is currently manufactured by Shell Chemical (Corterra Polymers) and DuPont (Sorona® 3GT). Based on its use in PTT synthesis, 1,3-propanediol has a potential 2020 market of 500 million pounds. The direct hydrogenation of 3HPA into 1,3-propanediol will require the development of new catalytic systems able to perform the direct reduction of the carboxylic acid groups to alcohol with good success. However, reduction of 3HPA esters to 1,3-propanediol is technically easier and can be performed efficiently using hydrogen as reducing agent in the presence of Cu based catalysts such as Cu/ZnO [45], or CuO/SiO$_2$ [46].

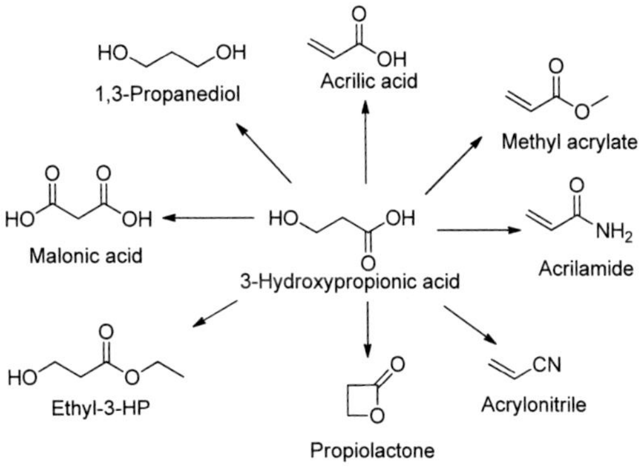

Scheme 3 3-Hydroxypropionic acid platform

The alcohol function of 3HAP can be dehydrated leading unsaturated compounds such as acrylic acid. Acrylic acid and derivatives (esters, salts or amides) are important compounds for the plastic industry. They are used as monomers in the manufacture of polymers and copolymers with numerous applications such as surface coatings, absorbents, textile papermaking, sealants, adhesives, etc. Biobased acrylic acid can be obtained in good yields by thermal dehydration in the liquid phase of 3HPA using homogeneous acid catalysis (sulfuric or phosphoric) in the presence of copper powder as a polymerization inhibitor [47]. Heterogeneous catalysts have also been used for dehydration of 3HPA to acrylic acid. For instance, excellent yield (96 %) of acrylic acid can be obtained at 180 °C, using NaH_2PO_4 supported on silica gel as catalyst, while other catalysts such as H-Beta zeolite [48], γ-Al_2O_3, SiO_2 and TiO_2 [49] have been used with different success.

3HPA can also be oxidized to malonic acid. Malonic acid and its esters are important intermediates in the synthesis of vitamins B_1 and B_6, barbiturates as well as other pharmaceuticals, agrochemicals and flavor and fragrance compounds. In the patent literature there are few examples concerning the catalytic oxidation of 3HPA. Particularly, it has been performed using oxygen as oxidant and Pt or Pd supported on C or Al_2O_3 as heterogeneous catalyst [50].

2. 2. 4 Levulinic acid

Levulinic acid (LEA) or 4-oxo-pentanoic acid is an useful intermediate for the synthesis of pharmaceuticals and solvents. Moreover its potential applications as plasticizer, coating materials and antifreeze compound have been reported [51]. Also, it has been described that LEA confers radiation protective properties to polymers and its esters are used as additive to increase thermal stability of poly (vinylchloride) [52].

LEA is obtained by dehydration of hexoses at moderate temperature under aqueous acid conditions leading to HMF, which is subsequently hydrated to obtain an equimolar mixture of levulinic acid and formic acid along with larger amounts of solid humic acids as side products. At industrial scale levulinic acid is produced by acid treatment of wood containing cellulose or hemicellulose. The process starts by the hydrolysis of polysaccharides to monosaccharides, the dehydration to obtain HMF and finally the hydration in the presence of strong acids such as sulphuric acid and hydrogen chloride. In this process about 40% yield of levulinic acid with

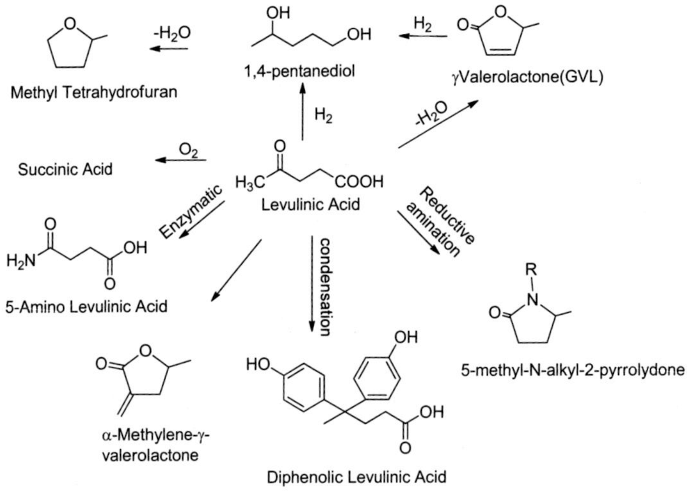

Scheme 4 Levulinic acid platform

respect to hexose content is obtained [53]. BioMetics has developed the Biofine process [54] which achieves industrial production of LEA from inexpensive lignocelluloses wastes including, agricultural residues, and wood waste and urban waste paper. In this processes LA is obtained with a theoretical maximum yield between 60 and 70% based on the hexose content. However this maximum yields is unlike to be reached because 1/3 of the carbon in the initial sugar is converted in unwanted black insoluble materials named humins. Generally these materials are burnt to generate energy (electricity and heat) for the process.

Hemicellulose derived pentoses via dehydration to furfural followed by hydrogenation to obtain furfuryl alcohol could also be a process to obtain LEA in high yield (95 %) [55]. (Scheme 4).

LEA has been of interest for many years because can be converted into valuable chemicals (Scheme 4). Thus, the reduction of levulinic acid gives 4-hydroxypentanoic acid which dehydrates to γ-Valerolactone (GVL). This compound can be hydrogenated to 1,4-pentanediol which can be dehydrated to form 2-methyltetrahydrofuran (see furan section).

Recently, there are great interests for the conversion of LA into liquid fuels. The key intermediate which allows efficient transformation involves the preparation of γ-Valerolactone (GVL). This biomass derived compound has numerous applications for making polymer fibers, as a solvent for lacquers, insecticides and food additive, and it is a potential biofuel for replacing ethanol in gasoline-ethanol blends [56]. GVL is typically obtained by catalytic hydrogenation of aqueous low temperatures (< 200 °C) using no acidic catalyst such as Ru/C, through the intermediate 4-hydroxypentanoic acid [57].

Interestingly, the formic acid formed during the sugar dehydration process can serve as a hydrogen source in the presence of ruthenium catalyst for LEA reduction to GVL, thereby avoiding the use of external fossil-fuel derived hydrogen [58].

2.2.5 Glycerol

Glycerol (1,2,3-propanetriol) is a high boiling point (290 °C), water-soluble alcohol that constitutes the backbone of triglycerides, which contribute approximately to 10% of total biomass. Glycerol is obtained as byproduct in the production of biodiesel. Transesterification of vegetable oils and animal fats with methanol in the presence of base catalysts, give fatty acid methyl esters (biodiesel) and glycerol. In this process, for every tone of biodiesel produced, 100kg of glycerol is generated. The rapid grow of biodiesel industry is generating a surplus of crude glycerol in the market which resulted in a significant fall of price for this feedstock (the price dropped ten times from 2005 to 2006). Nowadays, 50% of glycerol applications are food (as humectant and sweetener), health care and pharmaceuticals, nevertheless there are other markets for which crude glycerol may be suitable including the production of polyols, liquid detergents, renewable plastics and resins, for which the quality and purity requirements are generally lower [59]. Moreover, the disposal of abundant and cheap glycerol has generated in recent years a significant research effort to develop processes for their conversion into higher value chemicals.

Glycerol exhibits a versatile and high chemical reactivity due to its three hydroxyl groups. Some of the most important routes to obtain biodiesel derived glycerol are described below (Scheme 5).

Glycerol has traditionally played a role in the production of several types of polymers, some of which are available commercially. Polyglycerol (PG) and polyglycerol esters are used as biodegradable surfactants, lubricants, cosmetics, food additives, etc [60]. Moreover, glycerol based polymers offer additional benefits including lower amounts of leachability into the environment [61]. This application would compete with the more widely used polyethylenglycols. Polyglycerols are obtained from the oligomerization of glycerol and polyglycerol esters from the esterification or transesterification of the oligomers with fatty acids or methyl esters. Conventional methods for polymerizing glycerol require drastic process conditions, including high temperature and caustic or alkaline content. So it is not surprising that, with such a non selective process, a complex mixture of glycerol derivates is obtained. The reaction has been studied as well using either solid catalyst such as zeolites [62] or alkaline catalyst such as calcium hydroxide or sodium carbonate [63].

Aqueous glycerol can be gasified to CO, H_2 and CO_2 by a reforming process using metal-based catalysts [64]. The high reactivity of glycerol allows the reforming process to be carried out at milder temperatures (200–280°C). The gas mixture (CO + H_2) obtained (syngas) is a valuable starting material for methanol synthesis or the Fischer-Tropsch process [65].

Catalytic and thermal dehydration of glycerol provides several derivates. Depending on the conditions employed, dehydration occurs via loss of secondary hydroxyl group, leading to hydroxyacetone (acetol) which can be hydrogenated giving propylene glycol (1,2-PG or 1,2-propanediol). The reaction usually involves a bimetallic catalyst, such as copper-chromite material [66]. 1,2-PG, is a commodity chemical, with huge market in different sectors. When the glycerol dehydration occurs by a loss of a primary alcohol leads to the formation of 3-hydropropionaldehyde (3-HPA). 3-HPA can be hydrogenated to 1,3-propanediol (1,3-PDO) which is an important diol in the production of polyesters, polycarbonates and polyurethanes. Three examples of the large scale applications are Sorona®, Hytrel® made by Dupont and Corterra®, made by Shell. 1,3-PDO can be also dehydrated to acrolein, a precursor of acrylic acid, under high acidic conditions e.g., mineral acids such as H_2SO_4 or strong solids acids including

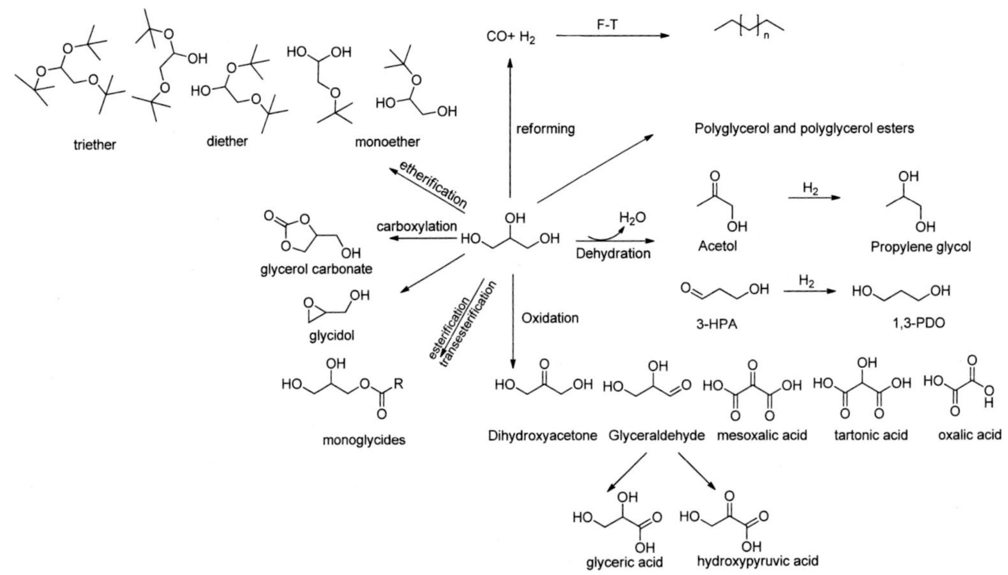

Scheme 5 Glycerol platform

ZrO$_2$-WO$_3$ [67] or zeolites [68]. As glycerol is already a highly functionalized molecule compared to hydrocarbons, an advantageous alternative is to use it as feedstock for the production of valuable oxygenated derivates. The oxidation of glycerol leads to a complex reaction pathway in which a large number of products with a wide range of applications such as dihydroxyacetone (DHA), glyceric acid (GLYAC), hydroxypiruvic acid (HPA), mesooxalic acid (MESAC), tartronic acid (TARAC), etc. could be obtained. To date, these products have limited market because they are either produced using costly and polluting stoichiometric oxidation processes. However, is possible to conduct the heterogenous catalytic oxidative reaction using inexpensive clean oxidizing agents such as air, oxygen, and hydrogen peroxide, instead of costly stoichiometric oxidants leading to an environmentally friendly alternative. Selective versus non selective oxidation is, therefore a big challenge associated with these catalytic oxidation reactions. An increasing number of studies dealing with the chemoselective catalytic oxidation of glycerol, mainly using supported noble metal nanoparticles such as Pd, Pt or Au have been reported recently. These studies showed that the nature of the metal as well as the pH of the medium controls the selectivity of the oxidation towards the primary or secondary alcohols [69]. It is out of the scope of these book chapter mention all the catalytic process that have been developed for the glycerol oxidation, however, as an example, over 1% Au/Charcoal or 1% Au/graphite catalyst, 100% selectivity towards glyceric acid can be obtained by the oxidation of glycerol under mild conditions [70].

Glycerol carbonate (GC) is another interesting material in the chemical industry which is employed mainly as solvent and in the surfactant industry. Inexpensive glycerol carbonate could serve as a source of new polymeric materials for the production of polycarbonates and polyurethanes. Traditionally, cyclic carbonates have been prepared by reaction of glycols with phosgene [71], but due to the high toxicity and corrosive nature of phosgene, greener alternative routes have been developed. For instance, the transesterification reaction of dialkyl or alkykene carbonates with glycerol using heterogeneous base catalysts [72] or by reaction of glycerol with urea using solid acid-base catalysts [73].

The fatty acid monoesters of glycerol are valuable compounds with wide applications as emulsifiers in food, pharmaceutical and cosmetic industry. Its detergent characteristics are due to its hydrophobic tail and its hydrophilic head [74]. The conventional synthesis of monoglycerides involves the transesterification of glycerol with vegetable oils (triglycerides) or with fatty acid methyl esters (biodiesel) using homogeneous base catalysts. The reaction products are usually a mixture of mono, di, and triesters with a monoglyceride content between 40–60% however improved results can be obtained using heterogeneous base catalysts [75].

Glycerol could also be processed though selective etherification into more valuable fuel additives or solvents with suitable properties. It can be converted into branched oxygen-containing components by catalytic etherification with either alcohols or alkenes. Among them tert-butyl ethers have potential as diesel fuel additives in gasoline and offer an alternative to methyl tert-butyl ether (MTBE) with is currently added to the fuels. For that reason is expected that this market will grow in the near future [76].

Another interesting glycerol derivative is glycidol. Due to its two functional groups, the alcohol and the epoxide, glycidol could be used in many organic syntheses. Glycidol can be used as well as a demulsifier, and dye-leveling agent. Thus it finds applications in surface coatings, chemical synthesis, pharmaceuticals, etc. It is generally produced by epoxidation of the allyl alcohol, however this process requires neutralization, extraction and purification due to the side products generated. A more friendly methodology starting from glycerol carbonate have been developed [77].

2. 2. 6 Furans

Monosaccharides such as glucose, fructose and xylose are the most available biomass primary compounds, which can be obtained from hydrolysis of cellulose, hemicellulose or starches. Dehydration of hexoses (glucose, fructose and mannose) allows obtaining 5-hydroxymethyl furfural (HMF) while dehydration of pentoses such as xylose leads to furfural. (Scheme 6). In this section we will give a general overview of the chemistry and catalytic transformations of

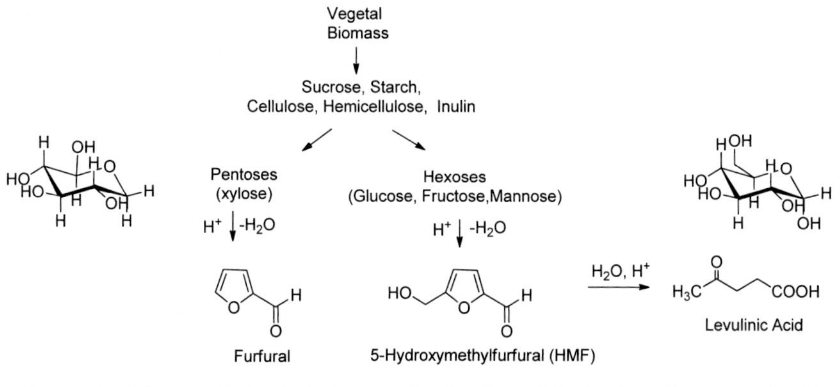

Scheme 6 Biomass-derived platform molecules from dehydration of monosaccharides

these furan derivatives.

2.2.7 5-Hydroxymethyl furfural (HMF)

5-Hydroxymethyl furfural (HMF) is an important biomass derived compound since it is a significant intermediate for the synthesis of a wide variety of chemicals and alternative biofuels [4a, 78].

HMF is prepared by triple dehydration of hexoses using mineral acids (H_2SO_4, HCl, H_3PO_4) [79]. However, the use of homogeneous catalysts has some drawbacks such as the need of a neutralization step, difficult catalyst recovering, use of expensive reactors resistant to corrosion etc. Owing to this, recent studies have been focussed on the use of heterogeneous catalysts with acidic properties which are easily separable and recyclable from the reaction medium, allowing to work at higher reaction temperatures and which in addition can be synthesized in a wide range of acidities allowing to improve the selectivity to HMF. In this context, a variety of heterogeneous acid catalysts including acid zeolites, vanadyl and niobium phosphates, ion exchange resins [80], vanadyl phosphates ($VOPO_4$) [81] etc., have been tested in the dehydration of sugars to HMF.

The selection of an appropriated reaction media is of paramount importance to achieve high selectivity in the dehydration of hexoses into HMF [82]. Thus, although aqueous acidic conditions are the most suitable from an ecological point of view, they are generally non-selective and yields of HMF are low due to the formation of levulinic acid, formic acid and insoluble humins [83]. On the other hand, biphasic systems (water-organic solvent) solve the problem arising from the low solubility of sugars in organic solvents, while the continuous extraction of the HMF from the aqueous phase prevents its degradation and avoid the formation of soluble polymers or humins. For example, 74% yield of HMF was obtained using a mixture of methyl isobutyl ketone (MIK) and water in the presence of a dealuminated acid zeolite (Mordenite) as catalyst [84]. Also Dumesic et al. have carried out the dehydration of fructose catalyzed by HCl in a water solution of DMSO in the presence of n-butanol and MIBK affording 89% selectivity to HMF at 95% of conversion [85]. In addition, high-boiling polar organic solvents, such as dimethyl sulfoxide (DMSO) [86], N-methylpyrrolidinone [87] and more recently ionic liquids [88] have been used in the HMF synthesis with good success. The advantage of using these solvents is that they avoid the formation of levulinic acid and humins, however they are difficult to separate from HMF, and in the case of ionic liquids their potential sensitive moisture, the cost for commercial applications and the need of purify after recycling, are the main disadvantages.

The dehydration of hexoses to HMF is more efficient and selective starting from fructose (a ketohexose) than from glucose (an aldohexose). However, glucose is still preferred in industry as starting material for HMF production because is more abundant and cheaper than fructose. Thus, current technologies employ an additional glucose isomerization step to fructose, which is subsequently dehydrated to HMF. For instance, using a biphasic system (H_2O/THF/NaCl) a Sn-Beta zeolite catalyst is able to isomerizes glucose to fructose which is subsequently dehydrated to HMF by a mineral acid (HCl) yielding HMF with 70% selectivity at 76% conversion [89].

A more complicate process to obtain HMF from disaccharides has been proposed by Takagaki [90]. using a physical mixture of a solid acid (Amberlyst-15) and a solid base (Hydrotalcite) as catalysts. Thus, disaccharides such as sucrose and cellobiose, were directly converted into HMF with high selectivity by sequential steps which include the hydrolysis of disaccharides by the acid catalyst, isomerization of glucose to fructose by the base catalyst and finally the dehydration of fructose by the acid sites. In the case of sucrose 93% selectivity to HMF at 58% conversion was achieved. Meanwhile a 99% of conversion and 76% selectivity to HMF were obtained starting from fructose.

The chemical structure of HMF, namely furan, primary hydroxyl and formyl functionalities allows synthetic transformations such as oxidation, reduction, condensation, etherification, etc., to other target molecules with multiple applications, e.g. pharmaceuticals, antifungals, and polymer precursors [4a, 82]. Some transformations of HMF into valuable biomass derivatives are presented in Scheme 7.

2,5-Furandicarboxylic acid (FDCA) and their dimethyl ester (DMFCA) are compounds with high potential applications in the polymer fields due to it can replace terephthalic, isoterephthalic and adipic acids, monomers widely used in the manufacture of polyesters, polyamides and polyurethanes [91]. It is known that polyethylene terephthalate (PET) is widely employed in the manufacture of fibers and plastic bottles, which is prepared by esterification reaction of ethylene glycol with terephthalic acid. In this sense, the substitution of terephathalic acid by 2,5-furandicarboxylic acid or its dimethyl ester which possesses similar structure and properties will allow the preparation of Bio-PET. Thus, HMF could find market in the Bio-PET production being a green alternative for petroleum derived terephthalic acid.

Generally FDCA is obtained by oxidation of HMF using conventional oxidants such as Ag_2O, $KMnO_4$ HNO_3 etc., giving yields between 47–70%. However, recently more environmental friendly oxidation systems which involve the use of air or oxygen as oxidants, heterogeneous metal catalysts, and controlled basic pH have been developed. For instance, gold supported on TiO_2 (Au/TiO_2) [92], or on nanoparticulated ceria (Au/CeO_2) [93] perform the oxidation of HMF into FDA in aqueous basic media in almost quantitative yields. On the other hand, when reactions are performed in methanol as a solvent, instead of basic water, dimethyl furan-2,5-dicarboxylate (DMFCA) is obtained through the one-pot oxidative esterification of HMF [94]. Moreover, by selecting the adequate catalyst, the oxidation of HMF to FDCA can be performed at neutral pH. Thus, Lilga et al [95]. have patented a method to oxidize quantitatively HMF to FDCA using a Pt/ZrO_2 catalyst and air as oxidant.

Another interesting HMF derivative is 2,5-diformylfuran (DFF) which is a versatile compound with various useful applications such as monomer [91a, 96] and as starting materials for the synthesis of pharmaceuticals [97] and antifungal agents [98].

Among the different catalytic systems investigated for the selective oxidation of HMF into DFF, vanadium based catalysts using air or oxygen as oxidants have been the most effective and selective [80b, 99].

Additionally, the synthesis of DFF directly from fructose (in one-pot reaction) or in two steps has been described by Carlini [100]. Thus, fructose is converted to HMF and then oxidized to DFF using vanadyl phosphate as catalyst and DMSO as a solvent under air pressure. Conversion of 84% of HMF with 97% of selectivity to DFF was obtained.

Selective reduction of the formyl group of HMF allows the formation of 2,5-bis (hydroxymethyl)

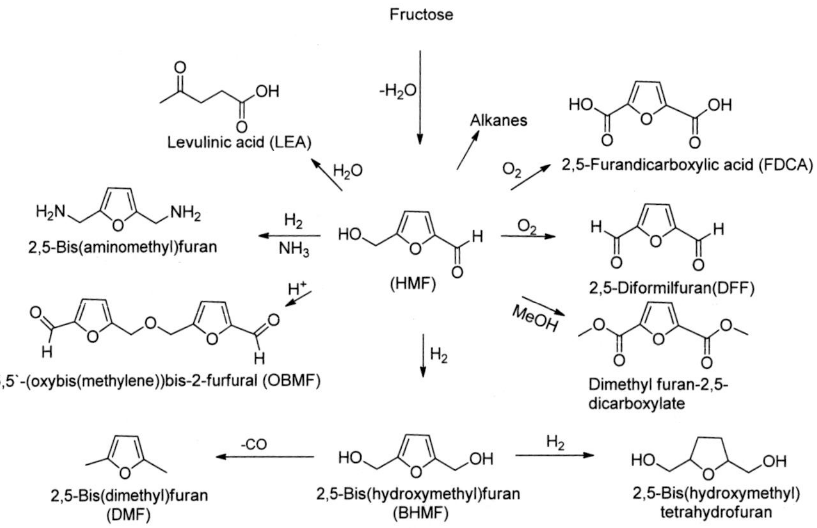

Scheme 7 Transformation of HMF into chemicals

furan (BHMF) a valuable product in the production of polymers and polyurethane foams [101], crown ethers [102] as an intermediate in the synthesis of drugs as well as solvents [103].

Further hydrogenation of BHMF leads to 2,5-bis (hydroxymethyl) tetrahydrofuran, which is used for the manufacture of polyesters. BHMF is mainly obtained by catalytic hydrogenation of HMF in aqueous phase in the presence of copper chromite [104] and platinum catalysts [105], while the use of nickel or palladium based catalysts (Pd/C) give preferentially the fully hydrogenated furan ring 2,5-bis (hydroxymethyl) tetrahydrofuran. In both cases high temperature (140–200 °C) and hydrogen pressures (70–75 bar) are required.

Additionally, 2,5-Dimethylfuran (DMF), a hydrophobic compound which possesses excellent characteristics such as high energy content and boiling point to be used as biofuel, can be obtained from HMF deoxygenation. The reaction involves hydrogenation of formyl group and subsequent C-O hydrogenolysis of the BHMF (see Scheme 7) using Cu based catalysts [78].

5,5'-(Oxy-bis (methylene)) bis-2-furfural (OBMF) is an interesting HMF derivative for the preparation of imine based polymers [103, 106]. It can be obtained by Williamson etherification reaction of HMF with 5-chloromethyl 2-furfural in the presence of an excess of homogeneous base or by etherification of HMF catalyzed by homogeneous acid catalysts such as *p*-toluenesulfonic acid [107]. In both cases the yields of OBMF do not exceed 72%. However, recently has been reported that micro- and mesoporous aluminosilicates bearing Bronsted or Lewis acid sites [108] are able to perform the etherification of HMF to OBMF with excellent success. For instance, using a mesoporous aluminosilicate (Al-MCM-41), a 99% yield of OBMF could be obtained in only 1 h reaction time.

2. 2. 8 Furfural

Furfural (furan-2-carboxaldehyde) with a production volume of 2.5×10^5 tons / year is one of the most common chemicals derived from lignocellulose. It is produced by dehydration of pentoses (xylose) and is a key derivative for the production of important non-petroleum-derived chemicals competing with crude oil.

The commercial process for the production of furfural, Quaker Oats technology, is based on the use of concentrated sulphuric acid as catalyst at moderate temperatures (< 200 °C). In this way, one-step hemicellulose hydrolysis followed by xylose dehydration yield furfural (less than 50%). Recent studies have been focused on the replacement of mineral acids by non corrosive, non toxic, stable and recyclable acidic solids including zeolites (HY, Mordenite) [109], microporous and mesoporous niobium silicalites [110], layered titanates, niobates and titanoniobates (HTiNbO$_5$) [111], among others for the dehydration of xylose to furfural. Reactions are usually carried out under aqueous-organic biphasic solvent system composed of water and toluene (W/T) under batchwise processing. In this system, the xylose conversion proceeds predominantly in the aqueous phase where it is completely dissolved and the furfural formed, which has a higher affinity for toluene, can be separated continuously from xylose and polar intermediates, avoiding losses by overreaction [112].

Extensive revisions about the chemistry of furfural and its derivatives can be found in literature [55, 113]. Therefore, we will describe here some of the main important transformations of furfural into chemicals related with the polymer industry and for production of fuels (Scheme 8).

The majority of furfural production is converted to furfuryl alcohol which is the most important furfural derivative with a variety of industrial applications. It is mainly used in the manufacture of resins, as starting material for the production of tetrahydrofurfuryl alcohol, and as chemical intermediate in the selective hydrogenation the synthesis of vitamin C, lysine and fragrances [114]. It is prepared industrially by the catalytic hydrogenation of furfural in liquid or vapour phase over copper chromite catalyst with good success [115]. A number of studies have been focused to find more environmentally and acceptable catalysts. Thus, for instance, platinum deposited on monolayer supports (SiO$_2$, MgO, TiO$_2$, Al$_2$O$_3$) [116], Cu/MgO, or Cu/La supported on MCM-41 [117] are able to catalyze the selective hydrogenation of furfural to furfuryl alcohol in yields higher than 94%. Recently Stevens et al [118]. have reported that by using supercritical CO$_2$ and two consecutive fixed flow reactors loaded with copper chromite and Pd/C respectively, furfural can be selectively hydrogenated to furfuryl alcohol (97% yield), tetrahydrofurfuryl alcohol (96% yield), 2-methylfuran (90% yield), 2-methyltetrahydrofuran (82% yield), and furan (98% yield) by controlling hydrogen concentration and reaction temperatures.

Additionally, furfuryl alcohol can be utilized for the production of levulinic acid, an important biomass platform molecule presented in the above section. Thus, the hydrolysis of furfuryl alcohol under aqueous acidic solutions at moderate temperature yields 80% of levulinic acid [52].

Complete hydrogenation of the furanic ring of furfural can be performed at high hydrogen pressures (> 20 bar) in the presence of supported noble metal as catalysts to produce methyltetrahydrofuran (MTHF) [65]. This apolar compound can be blended with gasoline up to 60% (v/v) without adverse effects on engine performances. Furthermore, it can be used as an excellent substitute of dichloromethane, a solvent widely used in the manufacture of pharmaceuticals, due to its superior properties such as better extraction characteristics for polar compounds, higher

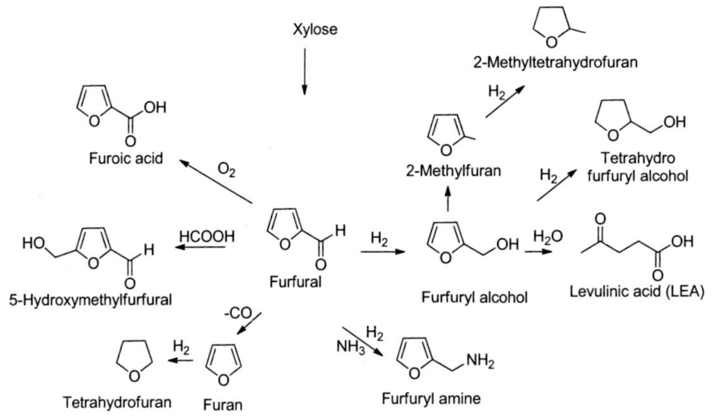

Scheme 8 Furfural as a platform molecule for the production of chemicals

boiling point (80 °C) and less toxicity.

Furfural is also the key intermediate for the commercial production of furan and its derivative, tetrahydrofuran (THF). Furan is obtained by catalytic decarbonylation of furfural using Pd supported on basic supports as catalysts which can be completely reduced to THF [119].

Furfurylamine is another interesting furfural derivative used in the manufacture of fibers, pharmaceuticals, and pesticides. It is industrially prepared by the reductive amination of furfural. Thus when furfural is reacted with NH_3 and hydrogen in the presence of Co and/or Ni based catalyst furfurylamine is obtained in 97% yield [120].

Furfural can also serve as a good source of carboxylic acids (such as furoic and maleic acids) *via* oxidation. Furoic acid is an important feedstock in the synthesis of pharmaceuticals and agrochemicals. Different conventional oxidants such as $KMnO_4$ have been utilized to oxidize furfural to furoic acid. However more recently full conversion of furfural to furoic acid has been performed using oxygen as oxidant in the presence of a bimetallic lead/platinum deposited on charcoal catalyst under basic aqueous solutions [121].

Finally, it is interesting to point out that furfural, can be transformed into hydrocarbons with molecular weights (C_8-C_{13}) appropriate for diesel and jet fuel applications by a cascade process which includes dehydration, hydrogenation and aldol condensation reactions [122].

References

1. Dodds, D. R., Gross, R. A., *Science*, **318**, 1250–1251 (2007)
2. Werpy, T. P., G. "Top Value Added Chemicals from Biomass: vol. I, Results of Screening for Potential Candidates from Sugars and Synthesis gas," U.S department of Energy Efficiency and Renewable Energy (2004)
3. Bozell, J. J., Petersen, G. R., *Green Chem.*, **12**, 539–554 (2010)
4. a) Corma, A., Iborra, S., Velty, A., *Chem. Rev.*, **107**, 2411–2502 (2007); b) Gallezot, P., *Chem. Soc. Rev.*, **41**, 1538–1558 (2012); c) Gandini, A., *Green Chem.*, **13**, 1061–1083 (2011)

5. Datta, R., Henry, M., *J. Chem. Technol. Biotechnol.*, **81**, 1119–1129 (2006)
6. a) Min-tian, G., Hirata, M., Koide, M., Takanashi, H., Hano, T., *Process Biochem.*, **39**, 1903–1907 (2004); b) Wasewar, K. L., Yawalkar, A. A., Moulijn, J. A., Pangarkar, V. G., *Ind. Eng. Chem. Res.*, **43**, 5969–5982 (2004)
7. Holm, M. S., Saravanamurugan, S., Taarning, E., *Science*, **328**, 602–605 (2010)
8. Robert L, A., *Catal. Today.*, **37**, 419–440 (1997)
9. a) Amgoune, A., Thomas, C. M., Roisnel, T., Carpentier, J.-F., *Chem.--Eur. J.*, **12**, 169–179 (2006); b) Nijenhuis, A. J., Grijpma, D. W., Pennings, A. J., *Macromolecules.*, **25**, 6419–6424 (1992)
10. Weissermel, K. A., H, "*Industrial organic chemistry*", Weinheim (2003)
11. Holmen, R. E. US Patent 2859240 (1958)
12. Sawicki, R. A. US Patent 4729978 (1988)
13. Paparizos, C. S., W, Dolhyj, S. EP Patent 181718 (1985)
14. Walkup, P., Rormannm, C., Hallen, R. US Patent 5071754 (1994)
15. Takafumi, A. S., H. US Patent 5250729 (1993)
16. Danner, H., Ürmös, M., Gartner, M., Braun, R., *Applied Biochemistry and Biotechnology.*, **70–72**, 887–894 (1998)
17. Adkins, H., Billica, H. R., *J. Am. Chem. Soc.*, **70**, 3118–3120 (1948)
18. a) Cortright, R. D., Sanchez-Castillo, M., Dumesic, J. A., *Appl. Catal., B.*, **39**, 353–359 (2002); b) Luo, G., Yan, S., Qiao, M., Fan, K., *J. Mol. Catal. A: Chem.*, **230**, 69–77 (2005); c) Mao, B.-W., Cai, Z.-Z., Huang, M.-Y., Jiang, Y.-Y., *Polym. Adv. Technol.*, **14**, 278–281 (2003); d) Santiago, M. A. N., Sánchez-Castillo, M. A., Cortright, R. D., Dumesic, J. A., *J. Catal.*, **193**, 16–28 (2000); e) Zhang, Z., Jackson, J. E., Miller, D. J., *Appl. Catal., A.*, **219**, 89–98 (2001)
19. a) Il Choi, J., Hong, W. H., Chang, H. N., *Int. J. Chem. Kinet.*, **28**, 37–41 (1996), b) Narayana, K. US Patent 6342626 (2002)
20. Sanz, M. T., Murga, R., Beltrán, S., Cabezas, J. L., Coca, J., *Ind. Eng. Chem. Res.*, **41**, 512–517 (2002)
21. Carlos Serrano-Ruiz, J., Dumesic, J. A., *Green Chem.*, **11**, 1101–1104 (2009)
22. Wadley, D. C., Tam, M. S., Kokitkar, P. B., Jackson, J. E., Miller, D. J., *J. Catal.*, **165**, 162–171 (1997)
23. Sauer, M., Porro, D., Mattanovich, D., Branduardi, P., *Trends Biotechnol.*, **26**, 100–108 (2008)
24. Wendisch, V. F., Bott, M., Eikmanns, B. J., *Curr. Opin. Biotechnol.*, **9**, 268–274 (2006)
25. McKinlay, J., Vieille, C., Zeikus, J., *Appl. Microbiol. Biotechnol.*, **76**, 727–740 (2007)
26. Huh, Y. S., Jun, Y.-S., Hong, Y. K., Song, H., Lee, S. Y., Hong, W. H., *Process Biochem.*, **41**, 1461–1465 (2006)
27. Delhomme, C., Weuster-Botz, D., Kuhn, F. E., *Green Chem.*, **11**, 13–26 (2009)
28. a) Song, H., Lee, S. Y., *Enzyme Microb. Technol.*, **39**, 352–361 (2006); b) Bechthold, I., Bretz, K., Kabasci, S., Kopitzky, R., Springer, A., *Chem. Eng. Technol.*, **31**, 647–654 (2008)
29. Noordover, B. A. J., van Staalduinen, V. G., Duchateau, R., Koning, C. E., van, B., Mak, M., Heise, A., Frissen, A. E., van Haveren, J., *Biomacromolecules.*, **7**, 3406–3416 (2006)
30. Varadarajan, S., Miller, D. J., *Biotechnol. Prog.*, **15**, 845–854 (1999)
31. Cukalovic, A., Stevens, C. V., *Biofuels, Bioprod. Biorefin.*, **2**, 505–529 (2008)
32. a) Hollstein, E. J. US Patent 3812148 (1974); b) Matson, M. US Patent 4904804 (1990)
33. Budroni, G., Corma, A., *J. Catal.*, **257**, 403–408 (2008)
34. Fumagalli, C., Updated by, S. In *Kirk-Othmer Encyclopedia of Chemical Technology*, John Wiley & Sons, Inc. (2000)
35. Petrini, M., Ballini, R., Marcantoni, E., Rosini, G., *Synth. Commun.*, **18**, 847–853 (1988)
36. Z. H, Z., *J. Mol. Catal. A: Chem.*, **168**, 147–152 (2001)

37. Corma, A., Garcia, H., Iborra, S., Primo, J., *J. Catal.*, **120**, 78–87 (1989)
38. Olah, G. A., Keumi, T., Meidar, D., *Synthesis*, **1978**, 929, 930 (1978)
39. a) Budarin, V., Luque, R., Macquarrie, D. J., Clark, J. H., *Chem.--Eur. J.*, **13**, 6914–6919 (2007); b) Budarin, V. L., Clark, J. H., Luque, R., Macquarrie, D. J., Koutinas, A., Webb, C., *Green Chem.*, **9**, 992–995 (2007)
40. Wolf, W. J., Updated by, S. In *Kirk-Othmer Encyclopedia of Chemical Technology*, John Wiley & Sons, Inc. (2000)
41. Jiang, X., Meng, X., Xian, M., *Appl. Microbiol. Biotechnol.*, **82**, 995–1003 (2009)
42. Brown, S. F., *Fortune.* (2003)
43. Mochizuki, M., Hirami, M., *Polym. Adv. Technol.*, **8**, 203–209 (1997)
44. Kasuya, K.-i., Inoue, Y., Doi, Y., *Int. J. Biol. Macromol.*, **19**, 35–40 (1996)
45. Forschner, T. C. P., J.B, Slaugh, L.H, Weider, P.R. WO Patent 2000018712 (1999)
46. Lee, B. N. J., L.S, Jang, E.J, Lee, J.H, Kim, H.R, Han, Y. US Patent 2003069456 (2002)
47. Amberst, B. C. US Patent 2469701 (1946)
48. Tsobanakis, P. M., X, Abraham, T. WO Patent 2003082795 (2003)
49. Cracium, L. B., G, Dewing, J, Schriver, G, Peer, W, Siebenhaar, B, Siegrist, U. US Patent 2005222458 (2005)
50. a) Co, M. P. JP Patent 57042650 (1980); b) Hass, T., Meier, M., Brossmer, C., Arntz, D., Freund, A. DE Patent 19629372 (1996)
51. a) Ghorpade, V. M., A., H. M. (1996); b) Manzer, L. E., *Acs Sym Ser*, **921**, 40–51 (2006)
52. V. Timokhin, B., A. Baransky, V., D. Eliseeva, G., *Prog. Polym. Sci.*, **68**, (1999)
53. Sunjik V, J., H., B., K., *Kem. Ind.*, **33** 599 (1984)
54. FritzpatrickS.W. WO Patent 9640609 (1997)
55. Zeitsch, K. J., "*The Chemistry and Technology of Furfural and its many by-products*", Amsterdam (2000)
56. Bond, J. Q., Alonso, D. M., Wang, D., West, R. M., Dumesic, J. A., *Science*, **327**, 1110–1114 (2010)
57. Serrano-Ruiz, J. C., Wang, D., Dumesic, J. A., *Green Chem.*, **12**, 574–577 (2010)
58. Deng, L., Li, J., Lai, D. M., Fu, Y., Guo, Q. X., *Angew. Chem., Int. Ed.*, **48**, 6529–6532 (2009)
59. Pagliaro, M. R., M In *The Future of Glycerol: New Uses of a Versatile Raw Material*, The Royal Society of Chemistry (2008)
60. Clacens, J. M., Pouilloux, Y., Barrault, J., *Appl. Catal., A.*, **227**, 181–190 (2002)
61. Roussel, C., Marchetti, V., Lemor, A., Wozniak, E., Loubinoux, B., Gérardin, P., *Holzforschung*, **55**, 57 (2001)
62. a) Harris, E. G. H., U, Bunte, R, J. Hachgnei, Kuhm, P. US Patent 5349094; b) Eshuis, J. I. L., J.A, Potman, R.P. US Patent 5635588
63. Stuhler, H. US Patent 4551561
64. Luo, N., Fu, X., Cao, F., Xiao, T., Edwards, P. P., *Fuel.*, **87**, 3483–3489 (2008)
65. Serrano-Ruiz, J. C., Luque, R., Sepulveda-Escribano, A., *Chem. Soc. Rev.*, **40** (2011)
66. Dasari, M. A., Kiatsimkul, P.-P., Sutterlin, W. R., Suppes, G. J., *Appl. Catal., A.*, **281**, 225–231 (2005)
67. Ulgen, A., Hoelderich, W., *Catal. Lett.*, **131**, 122–128 (2009)
68. Kim, Y. T., Jung, K.-D., Park, E. D., *Microporous Mesoporous Mater.*, **131**, 28–36 (2010)
69. Wyman, C. E., *Biotechnol. Prog.*, **19**, 254–262 (2003)
70. Carrettin, S., McMorn, P., Johnston, P., Griffin, K., Kiely, C. J., Attard, G. A., Hutchings, G. J., *Top. Catal.*, **27**, 131–136 (2004)
71. Kawata, O. JP Patent 6009610 (1994)
72. a) Bell, J., Arthur, J. R. US Patent 2915529, 1959, b) Mouloungui, Z. Y., J.W, Gachen, C.L, Gaset,

A. EP Patent 0739888 (1996)
73. a) Climent, M. J., Corma, A., De Frutos, P., Iborra, S., Noy, M., Velty, A., Concepción, P., *J. Catal.*, **269**, 140–149 (2010); b) Okutsu, M. K., T. EP Patent 11566042 (2001)
74. Baumann, H., Bühler, M., Fochem, H., Hirsinger, F., Zoebelein, H., Falbe, J., *Angew. Chem., Int. Ed.*, **27**, 41–62 (1988)
75. a) Corma, A., Hamid, S. B. A., Iborra, S., Velty, A., *J. Catal.*, **234**, 340–347 (2005); b) Corma, A., Iborra, S., Miquel, S., Primo, J., *J. Catal.*, 173, 315–321 (1998)
76. Wessendorf, R., *Erdoel Kohle, Erdas Petrochem.*, **48**, 138–143 (1995)
77. Teles, J. H. US Patent 5359094 (1994)
78. Roman-Leshkov, Y., Barrett, C. J., Liu, Z. Y., Dumesic, J. A., *Nature*, **447**, 982-U985 (2007)
79. Asghari, F. S., Yoshida, H., *Carbohyd. Res.*, **341**, 2379 (2006)
80. a) Mercadier, D., Rigal, L., Gaset, A., Gorrichon, J. P., *J. Chem. Technol. Biotechnol.*, **31**, 497–502 (1981); b) Moreau, C., Durand, R., Pourcheron, C., Tichit, D., *Stud Surf Sci Catal*, **108**, 399–406 (1997)
81. Carlini, C., Patrono, P., Galletti, A. M. R., Sbrana, G., *Appl. Catal., A.*, **275**, 111–118 (2004)
82. Rosatella, A. A., Simeonov, S. P., Frade, R. F. M., Afonso, C. A. M., *Green Chem.*, **13**, 754–793 (2011)
83. Rapp, M. K. EP Patent 0230250 (1987)
84. Moreau, C., Durand, R., Razigade, S., Duhamet, J., Faugeras, P., Rivalier, P., Ros, P., Avignon, G., *Appl. Catal., A.*, **145**, 211–224 (1996)
85. Chheda, J. N., Roman-Leshkov, Y., Dumesic, J. A., *Green Chem.*, **9**, 342–350 (2007)
86. Musau, R. M., Munavu, R. M., *Biomass.*, **13**, 67–74 (1987)
87. Sanborn, A., Bloom, P. D. 2006063287 (2006)
88. Stahlberg, T., Sorensen, M. G., Riisager, A., *Green Chem.*, **12** (2010)
89. Nikolla, E., Roman-Leshkov, Y., Moliner, M., Davis, M. E., *ACS Catal.*, **1**, 408–410 (2011)
90. Takagaki, A., Ohara, M., Nishimura, S., Ebitani, K., *Chem. Commun.*, 6276–6278 (2009)
91. a) Gandini, A., Belgcaem, M. N., *Prog. Polym. Sci.*, **22**, 1203–1379 (1997); b) Moreau, C. B., N.M. Gandini, A, *Top Catal.*, **27** (2004)
92. Gorbanev, Y. Y., Klitgaard, S. K., Woodley, J. M., Christensen, C. H., Riisager, A., *ChemSusChem.*, **2**, 672–675 (2009)
93. Casanova, O., Iborra, S., Corma, A., *ChemSusChem.*, **2**, 1138–1144 (2009)
94. a) Taarning, E., Nielsen, I. S., Egeblad, K., Madsen, R., Christensen, C. H., *ChemSusChem.*, **1**, 75–78 (2008); b) Casanova, O., Iborra, S., Corma, A., *J. Catal.*, **265**, 109–116 (2009)
95. Lilga, M. A., Hallen, R. T., Hu, J., White, J. F., Gray, M. J. US 20080103318 (2008)
96. Gandini, A., Belgcaem, N. M., *Polymer International* **47**, 267 (1998)
97. Gauthier, D. R., Szumigala, R. H., Dormer, P. G., Armstrong, J. D., Volante, R. P., Reider, P. J., *Org. Lett.*, **4**, 375–378 (2002)
98. Del Poeta, M., Schell, W. A., Dykstra, C. C., Jones, S., Tidwell, R. R., Czarny, A., Bajic, M., Bajic, M., Kumar, A., Boykin, D., Perfect, J. R., *Antimicrobial Agents and Chemotherapy*, **42**, 2495–2502 (1998)
99. Casanova, O., Corma, A., Iborra, S., *Top. Catal.*, **53**, 304 (2009)
100. Carlini, C., Patrono, P., Galletti, A. M. R., Sbrana, G., Zima, V., *Appl. Catal., A.*, **289**, 197–204 (2005)
101. Gandini, A. In *Agricultural and Synthetic Polymers*, American Chemical Society, Vol. 433 (1990)
102. Timko, J. M., Cram, D. J., *J. Am. Chem. Soc.*, **96**, 7159–7160 (1974)
103. Pentz, W. J. GB Patent 2131014 (1984)
104. Faury, A., Gaset, A., Gorrichon, J. P., *Inf. Chim.*, 214 (1981)

105. Schiavo, V., Descotes, G., Mentech, J., *Bulletin de la Societe Chimique de France* (1991)
106. Moreau, C., Belgacem, M. N., Gandini, A., *Top. Catal.*, **27**, 11–30 (2004)
107. Chundury, D., Szmant, H. H., *Ind. Eng. Chem. Res.* **20**, 158–163 (1981)
108. Casanova, O., Iborra, S., Corma, A., *J. Catal.*, **275**, 236–242 (2010)
109. Dias, A. S., Lima, S., Brandao, P., Pillinger, M., Rocha, J., Valente, A. A., *Catalysis Letters*, **108**, 179–186 (2006)
110. Dias, A. S., Lima, S., Carriazo, D., Rives, V., Pillinger, M., Valente, A. A., *Journal of Catalysis*, **244**, 230–237 (2006)
111. Lima, S., Pillinger, M., Valente, A. A., *Catal Commun*, **9**, 2144–2148 (2008)
112. Weingarten, R., Cho, J., Conner, W. C., Huber, G. W., *Green Chem.*, **12**, 1423–1429 (2010)
113. Lange, J.-P., van der Heide, E., van Buijtenen, J., Price, R., *ChemSusChem.*, **5**, 150–166 (2012)
114. Kottice, R. H. In *Kirk-Othmer Encyclopedia of Chemical Technology*, 4th ed., John Wiley & Sons, Inc.: New York (1997)
115. Hinnekens, H. DE Patent 3425758 (1984)
116. Kijeński, J., Winiarek, P., Paryjczak, T., Lewicki, A., Mikołajska, A., *Applied Catalysis A: General*, **233**, 171–182 (2002)
117. Hao, X.-Y., Zhou, W., Wang, J.-W., Zhang, Y.-Q., Liu1, S., *Chem. Lett.*, **7** (2005)
118. Stevens, J. G., Bourne, R. A., Twigg, M. V., Poliakoff, M., *Angew. Chem., Int. Ed.*, **49**, 8856–8859 (2010)
119. Sitthisa, S., Resasco, D. E., *Catalysis Letters*, **141**, 784–791 (2011)
120. Ayusawa, T., Mori, S., Aoki, T., Hamana, R. JP patent 60146885 (1984)
121. Verdeguer, P., Merat, N., Rigal, L., Gaset, A., *J. Chem. Technol. Biotechnol.*, **61**, 97–102 (1994)
122. Corma, A., de la Torre, O., Renz, M., Villandier, N., *Angew Chem Int Edit*, **50**, 2375–2378 (2011)

Chapter 3
Bio-polyesters

3.1 Poly(lactic acid)
Hitomi Ohara

3.1.1 Lactic acid fermentation
Raw materials

Corn, beet, and sugarcane have been used as the raw materials for lactic acid fermentation; however, the technology to produce lactic acid from nonedible wood biomass should be developed in the near future. Plant cell walls, the major source reservoir of fixed carbon in nature, have three major polymers: cellulose, hemicellulose, and lignin. Hemicellulose includes glucans, mannans, and xylan. The amount of xylan varies in different plants, from as much as 35% of the dry weight of birch wood to as little as 7% in some gymnosperms. To control carbon dioxide exhaust entering the atmosphere, ethanol made from biomass via fermentation has already been used for automobile fuel. Because *Saccharomyces,* typical yeast, cannot utilize xylose, genetically engineered *Saccharomyces* [1–8] or *Escherichia coli* [9, 10] strains were developed to convert xylose contained in biomass to ethanol. Thus, if xylose is utilized as a carbon source for lactic acid fermentation, wood biomass can be used effectively.

Various pretreatment options are now available to fractionate, solubilize, hydrolyze and separate cellulose, hemicelluloses, and lignin components. These include enzyme hydrolysis and dilute acid treatment. Dilute acid treatment at high temperature generally hydrolyzes hemicellulose to simple sugars but produces furan which inhibits fermentation. Enzyme hydrolysis does not produce fermentative inhibitors; it solubilizes hemicellulose and degrades it to a mixture of xylooligosaccharides. These xylooligosaccharides can be processed by microorganisms into various chemicals [11–19]. A mixture of hydrolyzed cellulose and hemicellulose should be used as the raw material in industrial production, because wood biomass is hydrolyzed simultaneously.

Microorganism

There are two types of lactic acid bacteria: homo-type where 2 moles lactic acid are produced from 1 mole glucose and hetero-type where 1 mole lactic acid, 1 mole carbon dioxide, and 1 mole ethanol or acetic acid are produced from 1 mole glucose. The former has the metabolic pathway called "Embden-Meyerhof pathway" which causes no loss of carbon sources and the latter includes the pathway of pentose phosphate cycle (Figure 1). Therefore, for the effective production of lactic acid, it is necessary to select proper strain. Hitherto, since lactic acid

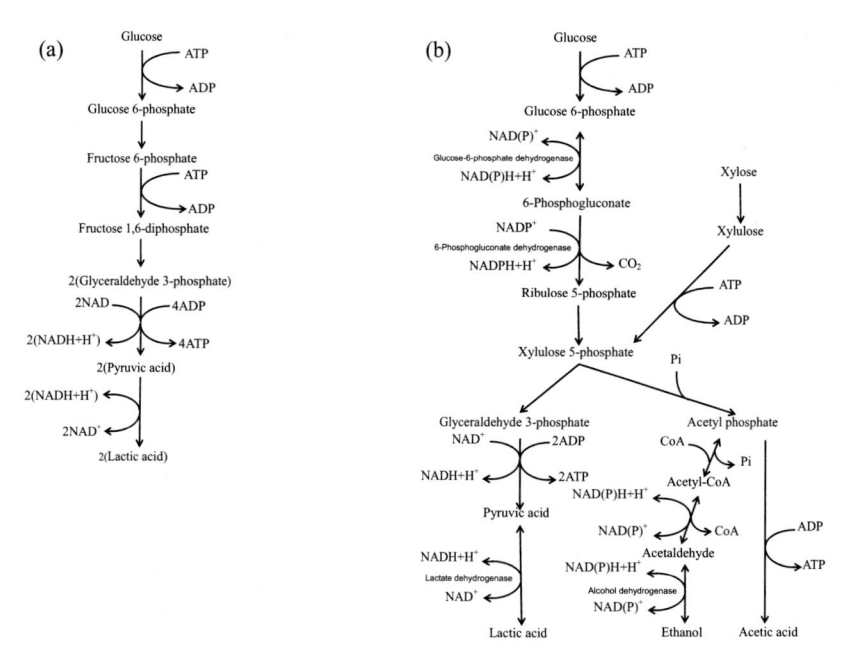

Figure 1 Two types of metabolic pathway of glucose to lactic acid
(a) Homo–type, (b) Hetero–type

fermentation has been mostly used in the food-related field, the research thereof has targeted on flavor and safety. However, in order to manufacture industrial lactic acid capable of serving as the raw material of poly(lactic acid), it is necessary to select microorganisms from a different perspective. Specifically, it is required to select homo-type microorganisms capable of generating neither carbon dioxide nor ethanol, of providing lactic acid in high optical purity and of less producing such by-products as are difficult to be removed by purification, for example, acetic acid and butyric acid. Further, it should be noted that the optical purity of lactic acid [20] and the metabolism of acetic acid [21, 22] may be varied according to the fermentation conditions. Lactic acid fermentation has been applied to foods since early times, and therefore the safety and purification of the microorganisms to be used have been studied from the viewpoint of their application to foods. Further, since WHO counseled that neither D-lactic acid nor D,L-lactic acid should be in infant foods [23], lactic acid manufactured now is mainly in the form of L-lactic acid. There is a report that investigates enantiomers of lactic acid produced by various microorganisms [24]. For example, Streptococci and some Lactobacilli are known as lactic acid bacteria capable of selectively producing L-lactic acid (Figure 2). However, microorganisms used in the manufacture of bio-based materials are required to be selected from a perspective different from food safety. For example, besides the above lactic acid bacteria, other lactic acid bacteria producing L-lactic acid of high optical purity are reported to be found in *Bacilli* of low auxotrophy [24, 25] and in *Rhizopus oryzae* [26]. Those bacteria of *Rhizopus* produce L-lactic acid only from saccharides and inorganic salts [27], but there is no report that the *Rhizopus* produces D-lactic acid. On the other hand, however, there are known microorganisms producing D-lactic acid. They are, for example, lactic acid bacteria such as *Lactobacillus delbrueckii* and *Sporolactobacillus inulinus*, and other microorganisms such as *Bacillus laevolacticus* [22]. In general, lactic acid bacteria are so highly auxotrophic that proteins and minerals as well as carbohydrates are indispensable. Accordingly, in some cases, those auxiliary materials rather than carbohydrates drive up the manufacturing cost. Further, because of large evaporation energy of water, it takes much energy to condense a lactic acid fermentation culture broth in the manufacturing process of lactic acid. The lactic acid fermentation culture broth is therefore preferably obtained in a concentration as high as possible, so as to reduce the manufacturing cost. If a buffer solution is used alone, lactic acid bacteria lose their activity about at pH 4.2, and the lactic acid concentration at that time is about 1.5 to 2.0%. In contrast, if pH-stat control is carried out, the lactic acid concentration becomes about 8 to 12%. Meanwhile, in the case where lactic acid is chemically synthesized, acetaldehyde as the raw material is subjected to an addition reaction with hydrogen cyanide. In Japan, lactic acid is synthesized by hydrolysis of lactonitrile, which is by-produced in manufacture of acrylonitrile [28] (Figure 3). However, those methods produce D,L-lactic acid. The optical purity thereof, which needs to be measured simply and accurately, can be determined by means of high performance liquid chromatography [20]. It is also possible to measure the optical purity by oxidation-reduction of NAD with lactate dehydrogenases, but this enzymatic method is too poor in precision to be used for lactic acid serving as the raw material of poly(lactic acid). That is because 1% or less of optical isomers must be detected in the optical purity measurement of lactic acid for poly(lactic acid).

In accordance with the recent progress of genetic engineering, it has been attempted to produce lactic acid by various microorganisms. For example, it is reported that a lactic acid-producing gene is transferred into *Saccharomyces cerevisiae*, which is a kind of yeast, to produce L-lactic acid [29-31]. Further, in these days, studies have been dynamically made to research microorganisms that produce D-lactic acid usable as a material of stereo-complex poly(lactic acid). For example,

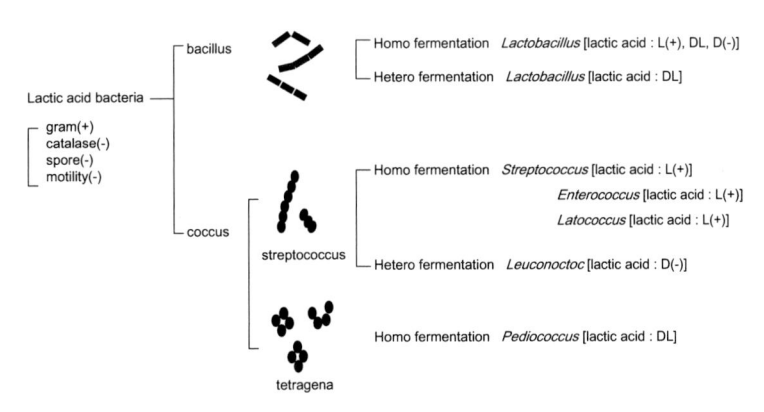

Figure 2 Morphology of Lactic acid bacteria

Figure 3 Chemical synthetic process of racemic lactic acid

it is reported that D-lactic acid is produced by gene recombination of *E. coli* [32]. It is also being studied that copolymer of lactic acid and hydroxybutyrate is produced in a microorganism. This copolymer contained lactic acid in 6 mol% at the beginning [33], but its content has increased gradually [34]. At present, the copolymer produced by this method has a weight average molecular weight of 2.0×10^4, which is not a very high value, but the molecular weight distribution thereof has such a narrow width that the PDI value is about 1.3.

Purification of lactic acid

No matter how favorably the fermentation may proceed, a raw material of poly(lactic acid) cannot be obtained in view of the quality and cost if wrong purification procedures are adopted. In lactic acid fermentation, lactic acid as the product decreases the pH value and consequently lowers the activity of bacteria. Accordingly, if the produced lactic acid is successively neutralized, a high-concentration lactic acid solution can be obtained. In fact, for controlling the pH value, calcium carbonate has been conventionally added into the fermentor. Since having low solubility to water, the added calcium carbonate precipitates in the fermentor. When lactic acid is produced, the produced acid is immediately neutralized to become calcium lactate and the precipitated calcium carbonate is consumed. After the remaining calcium carbonate is removed, the supernatant solution is concentrated to precipitate calcium lactate. The calcium lactate is collected and treated with sulfuric acid to be free lactic acid, which is then esterified with ethanol to prepare ethyl lactate. The ethyl lactate is purified by distillation and hydrolyzed to obtain purified lactic acid (Figure 4). In this process, sulfuric acid used for become free lactic acid from calcium lactate may remain in the subsequent step because it serves as a catalyst of esterification. Accordingly, the remaining sulfuric acid is favorable after all. However, this conventional process emits a large amount of gypsum as a by-product. Further, it is also a problem that carbon dioxide is generated from calcium carbonate by neutralization. Nowadays, during the lactic acid fermentation, the pH value is controlled by addition of NaOH or aqueous ammonia according as monitored with a steam-sterilizable pH meter. There is a report that studies a process adopting neutralization with ammonia [35]. Ammonium lactate obtained by the neutralized fermentation is converted into butyl lactate with Dean-Stark distillation apparatus, and the product is successively distilled with the apparatus [36]. The produced butyl lactate is hydrolyzed to obtain lactic acid (Figure 5). Unlike the conventional calcium method described above, this method does not emit a large amount of waste and further it is possible to recover ammonia used as a neutralizing agent. Meanwhile, it is said that some attempts are made to study what is called the direct distilling method, in which lactic acid is not converted into esters but directly distilled. However, in this method, lactic acid undergoes dehydration-polymerization when distilled, to form oligomers and consequently to lower the yield. Accordingly, it is necessary to take some measures against that. Further, it is also studied to convert lactates into free lactic acid by means of bipolar membranes or ion exchange resins [37], but there are many problems such as fouling of membranes and decrease of concentration.

Figure 4 Calcium method

Figure 5 Butanol ester method

3. 1. 2 Synthesis of poly(lactic acid)

The process of synthesizing poly(lactic acid) (PLA) from lactic acid can be roughly categorized into ring-opening polymerization process and direct polymerization process (Figure 6). The former is a process comprising the steps of: dimerizing lactic acid to form lactide (3,6-dimethyl-1,4-dioxane-2,5-dion), which is a cyclic dimer of lactic acid; and then subjecting the lactide to ring-opening polymerization. On the other hand, if lactic acid is simply subjected to polymerization, heating and dehydration-condensation, the obtained polymer has a weight average molecular weight of no more than about 3.0×10^3 [38]. However, in order to serve as materials of fibers, films and structure members, the polymer needs to have a weight average molecular weight of 1.5×10^5 to 2.0×10^5. Incidentally, even if the molecular weight becomes more than 2.0×10^5, the polymer scarcely changes its characteristics. Various processes, therefore, have been developed with aim of increasing the weight average molecular weight to 2.0×10^5.

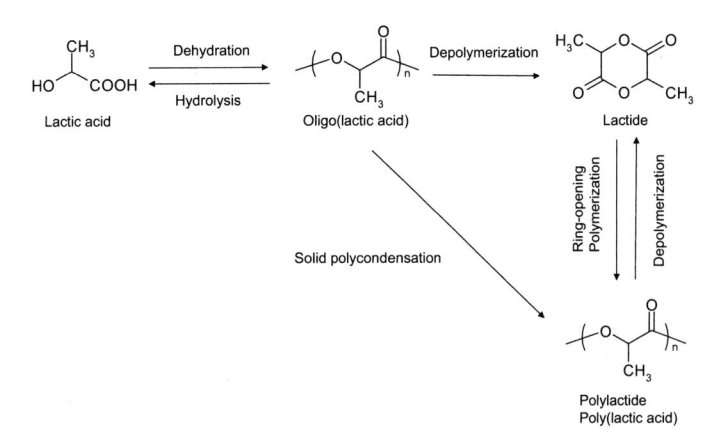

Figure 6 Synthetic routes of poly(lactic acid) and poly(lactide)

Lactide method

(1) Catalyst and reaction mechanism of ring-opening polymerization

Lactide readily undergoes ring-opening polymerization to form poly(lactic acid). This polymerization proceeds according to anionic or coordinated anionic mechanism [39–43], and catalyst most popularly used therein is tin 2-ethylhexanoate, which is experientially regarded as low toxic. In some cases, lauryl alcohol is added as an initiator of the ring-opening polymerization and the molecular weight of the resultant polymer is controlled by the amount thereof. However, in order to realize the polymerization degree corresponding to a weight average molecular weight of about 2.0×10^5, it is enough to use water adsorbed on the lactide crystallites as an initiator. Some reaction mechanisms have been proposed that describe the lactide polymerization in $Sn(Oct)_2$/ROH catalyst system. They are roughly categorized into the following four mechanisms: (i) cationic mechanism, with formation of secondary oxionium ions (*e.g.*, initiation with carboxylic acid) [44,45]; (ii) active monomer mechanism [46,47]; (iii) direct insertion mechanism [48, 49]; and (iv) coordination-insertion mechanism [50]. Figure 7 shows schemes of those mechanisms. Duda and Penczek revealed by TOF-MASS that a Sn atom is combined with the growing polymer-end (Sn-O-PLA), and accordingly the mechanism (iv) is now considered to be probable. According to the coordination-insertion mechanism, ligand exchange occurs between $Sn(Oct)_2$ and ROH to form -Sn-OR, which functions as an essential initiator so that the alkoxide makes a nucleophilic attack onto the carbonyl carbon of lactide activated with Sn to initiate the polymerization. Similarly, lactide is inserted into between Sn-oxygen to increase the polymerization degree in the growing

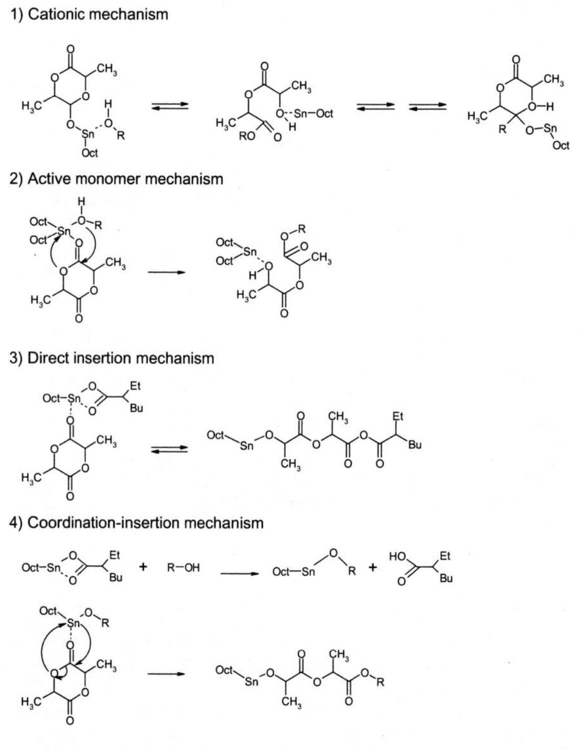

Figure 7 Proposed mechanisms for $Sn(Oct)_2$ to ring opening polymerization of lactide

reaction. On the other hand, assuming that the polymerization starts and proceeds according to the cationic mechanism, the Sn atom should not be combined with the growing polymer-end and further the cations serving as active species should be inactivated if basic amines are added as an initiator. Actually, however, the polymerization rate is improved by contraries when the amines are added. In addition, an octylic acid anion as the counter ion is not acknowledged. From those facts, the cationic mechanism is now regarded as improbable. Further, the activation energy at that time was calculated to be 20.6 kcal/mol by theoretical study [51]. Meanwhile, according to the active monomer mechanism, three species of $Sn(Oct)_2$, lactide and ROH form a complex to drive the polymerization. In this mechanism, the Sn atom is thought to participate in the reaction, from beginning to end, freely enough to be concerned with neither ionic nor covalent bond. However, this is inconsistent with the above-mentioned fact that Sn-O-PLA was detected. As for the direct insertion mechanism, the mechanism has been proposed based on the experimental result that $Sn(Oct)_2$ molecules were polymerized by themselves. However, no one has verified the end structure of acid anhydride generated by the nucleophilic attack of octylic acid anion onto the carbonyl carbon of the monomer.

When octylic acid is added into the $Sn(Oct)_2$/ROH system, the polymerization rate decreases according as the added amount increases. This means that, even if hydrolyzed in the presence of water to form octylic acid, $Sn(Oct)_2$ satisfyingly functions as a catalyst. However, if the polymerization is wanted to be well-controlled, it is said to be necessary to carefully dehydrate the whole reaction system. According to the coordination-insertion mechanism with Sn-OR formation, it is presumed that $Sn(Oct)_2$ by itself serves as neither a catalyst nor an initiator but Sn-OR serves as an essential initiator.

Other powerful catalysts are also usable, and examples of them include aluminum isopropoxide $(Al(O\text{-}iPr)_3)$, Zn^{2+}-type catalysts, and lanthanide-type catalysts $(Y(O\text{-}iPr)_3)$. Aluminum alkoxide complexes are easily available and hence widely used in various polymer systems. Accordingly, for polymerization of lactide, aluminum iso-propoxide $(Al(O\text{-}iPr)_3)$ is a typical catalyst. The $(Al(O\text{-}iPr)_3)$ is normally in such an aggregated form as an equilibrium mixture of liquid trimers and solid tetramers at room temperature. Figure 8 shows structures of the trimer and tetramer. After purified by distillation under reduced pressure, the equilibrium mixture can be directly used because having enough polymerizing activity. However, since it is a mixture, the polymerization caused thereby is difficult to be controlled and analyzed kinetically. Kowalski *et al.* therefore separated the trimer and the tetramer, and investigated each polymerization behavior in detail [52]. As a result, it was revealed that the initial reaction rate by the trimer is much faster than that by the tetramer, and it was also verified by NMR that the trimer readily reacts with the monomer to

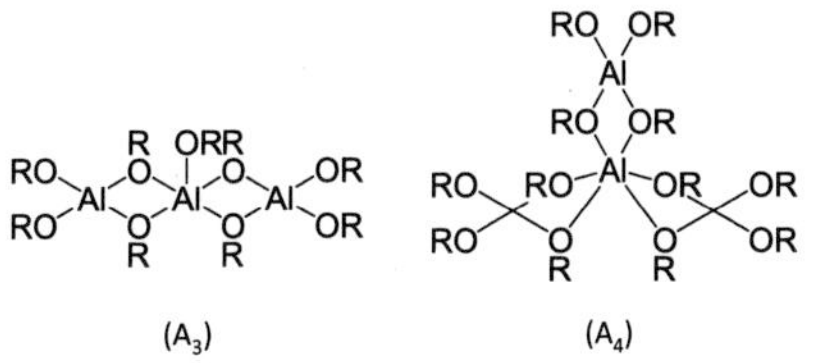

Figure 8 The structure of $Al(O\text{-}iPr)_3$ of trimer (A_3) and tetramer (A_4)

form an initiator species. Further, the trimer causes living polymerization, in which the molecular weight linearly increases in proportion to the monomer conversion, but the tetramer does not. However, when the monomer is almost completely consumed, both the trimer and the tetramer finally provide a polymer whose molecular weight is identical with the aimed molecular weight, which is determined by the feed ratio between the monomer and the initiator. The polymer given by the trimer alone or by the tetramer alone has a molecular distribution Mw/Mn (PDI) of 1.1 to 1.2 or 1.2 to 1.3, respectively, while the perfect living polymer has a PDI of almost 1. This suggests that ester-exchange reactions occur as side-reactions at the final stage of the polymerization and that they occur more frequently in the polymerization by the tetramer than by the trimer. The polymerization mechanism thereof is presumed to be a coordination-insertion mechanism in which the monomer is inserted into between Al-O, like into between Sn-OR described above, so as to grow evenly the polymer chain. Besides the above, there are many studies concerning alkoxide complex catalysts of transition metals such as Zn, Zr, Y, Fe, Ti and La [53].

However, since it has become regarded as a problem that metal catalysts used in lactide polymerization are toxic to living bodies, some proposals have been made for adopting oxygen polymerization and organic catalysts containing no transition metal. Matsumura *et al.* subjected lactide to bulk ring-opening polymerization with various kinds of lipase. As a result, they reported that a polymer having a weight average molecular weight (Mw) of 2.7×10^5 was obtained by polymerization of D,L-lactide with lipase PS [54, 55]. Judging from their supplier, the D,L-lactide used by them seems to be a racemic mixture of D,D-lactide and L,L-lactide. On the other hand, Keul *et al.* attempted to individually polymerize D,D-lactide and L,L-lactide with Novozyme 435, which is a kind of lipase. As a result, no polymer was detected from L,L-lactide, but a polymer having a number average molecular weight of 1.2×10^3 was obtained from D,D-lactide. The obtained polymer was a narrow distribution polymer. They presume that this polymerization is based on a reaction mechanism in which lactide is hydrolyzed with water and polymerized by catalysis of the lipase to form a straight-chain polymer of lactic acid [56]. If their presumption is correct, a polymer of higher molecular weight will be obtained by further reducing the water content to a very small amount.

Recently, Hedrick *et al.* have shown that heterocyclic carbene compounds, amidines and thiourea compounds have sufficient activity for lactide polymerization, that the polymerization is living, and that it can form a mono-dispersed polymer (Mw/Mn < 1.1) [57–59]. Figure 9 exhibits proposed reaction schemes of those catalysts. As for the carbene catalyst, it is thought that carbene makes a nucleophilic attack directly onto the carbonyl carbon of lactide to initiate the polymerization. In contrast, as for the amidine or thiourea catalyst, it is thought that two or more amino groups activate the monomer and the initiator (ROH) to initiate the polymerization. They mainly carried out solution polymerization characterized in that the reaction was completed in a short time (at least 20 seconds) at room temperature. The formed polymer had a number average molecular weight (Mn) of 6.0×10^4 at most. Further, according to their reports, carbene compounds in which the phenyl substituent groups of the hetero-rings are made more bulky exhibit stereo-selectivity at an extremely low temperature of 0°C or below [60, 61]. If *rac*- and *meso*-lactides are polymerized, the selectivity is improved with decrease of the polymerization temperature to

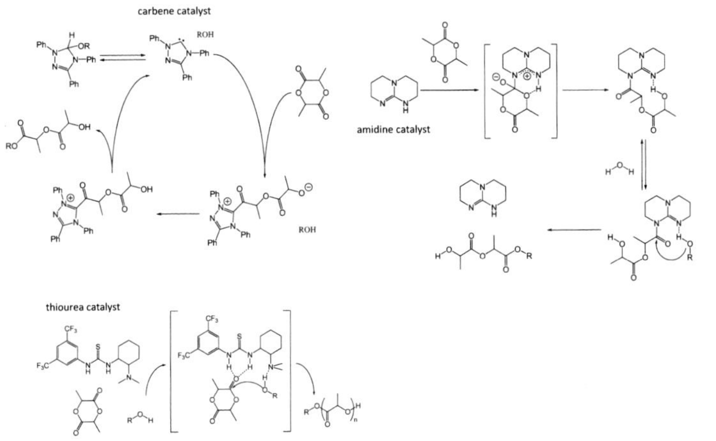

Figure 9 The various organic catalyst for the polymerization of lactide

obtain a stereo-block type or hetero-tactic type poly(lactic acid). Further, since the interaction with the monomer (cyclic ester) through hydrogen bonds differs in strength from that with the polymer (straight-chain ester), the amidine and thiourea catalysts can inhibit the ester-exchange reaction more effectively than other catalysts. This property seems to be advantageous for block polymerization. Li *et al.* have proposed various organic catalysts [62–64], and they verified that the bulk polymerization of lactide with a guanidium salt, which is a kind of the amidine catalyst, is living polymerization to produce a mono-dispersed polymer. That polymerization can be carried out under the same conditions as that in Sn(Oct)$_2$/ROH system. However, if those strong bases are used, there is fear of racemization of lactide.

Spassky *et al.* synthesized an Al complex having a ligand of SALBinapht, which is a chiral Schiff base, and found that its (R)-type complex preferentially polymerized D-lactide in *rac*-lactide [65]. In this polymerization, the ratio of D and L monomer consumption rates k_D/k_L was about 2.0, and therefore L-lactide started to be polymerized in the later stage of the reaction so as to introduce L-units gradually into a PDLA chain formed in the early stage. As a result, before the end of chain propagation, a PLLA chain was formed after the PDLA chain via atactic arrangement. Thus, the polymerization produced a block polymer referred to as "gradient PDLA-PLLA". It is considered that the stereo-selectivity of the catalyst is ascribed to two factors, namely, axial chirality of the ligand and bulkiness thereof. Baker and Smith *et al.* polymerized *rac*-lactide with a racemic mixture of the above complex, and consequently obtained an almost isotactic polymer having a melting point of 190°C. From this result, they presume that each of the (R) and (S) type complexes independently and selectively polymerized D- and L-lactides, respectively, and that the produced PLLA and PDLA chains formed the stereo-complex [66]. However, Cate *et al.* reviewed microstructures of the produced polymers and consequently concluded that each polymer was not a mixture of homopolymers but a block polymer of PLLA and PDLA. Further, since the catalysts had incomplete stereo-selectivity, they proposed a polymer chain-exchange mechanism and thought that the produced polymers had multi-block structures [67, 68]. Such PLLA-PDLA block-type poly (lactic acid) is referred to as "stereo-block poly(lactic acid) (sb-PLA)". According to their thought, the catalyst does not need to be chiral when *rac*-lactide is polymerized to form sb-PLA. In fact, there have recently been many reports in each of which sb-PLA is synthesized from *rac*-lactide with a complex having an achiral bulky ligand [69–71]. Figure 10 summarizes catalysts having stereo-selectivity. As described above, with respect to the stereo-selective polymerization of lactide, new catalysts are being developed one after another even now. As a result, it has become possible to synthesize novel poly(lactic acid) such as hetero-tactic type poly(lactic acid) (hetero-PLA) and syndiotactic type poly(lactic acid) (syn-PLA), which are shown in Figure 11 [72, 73]. There is possibility that those new kinds of PLA will be found to have characteristics different from conventional PLA, and therefore it is hoped that they will be further studied.

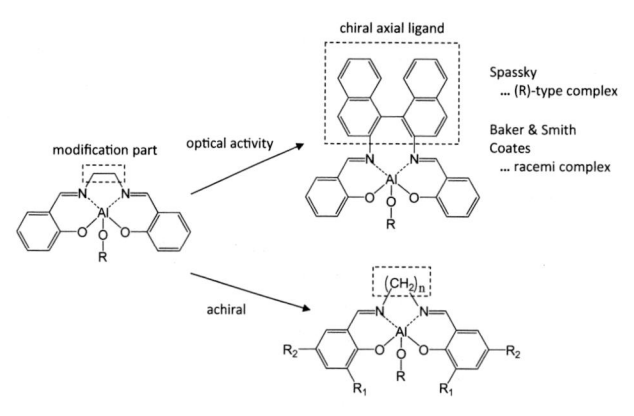

Figure 10 The various Al-salen complex suggested for the stereoselective polymerization of lactide

Figure 11 Lactide stereoisomer and poly(lactic acid) microstructure

3.1.3 Industrial manufacturing methods
Synthesis and purification of lactide
(1) Purified lactide method (Figure 12)

Lactic acid is commercially available normally in the form of 90% aqueous solution. first, the lactic acid aqueous solution is condensed and dehydrated by heating to synthesize lactic acid oligomer having a weight average molecular weight of 3.0×10^3 to 4.0×10^3. The lactic acid oligomer is then mixed with tin octylate as a catalyst, and heated under reduced pressure so as to distill out lactide. There is equilibrium between the oligomer and lactide, and the equilibrium leans toward the oligomer side at such a high temperature that lactide is synthesized in the presence of the catalyst. Accordingly, in order to advance the reaction, lactide is taken out of the reaction system by distillation in the reaction. During the distillation, dimer and trimer of straight-chain lactic acid are evaporated out together with lactide. Since those straight-chain lactic acid compounds inhibit ring-opening polymerization of lactide, purification is necessary. Lactide includes: L,L-lactide, which is a ring formed from two molecules of L-lactic acid; D,D-lactide, which is a ring formed from two molecules of D-lactic acid; and *meso*-lactide, which is a ring formed from L lactic acid and D-lactic acid. It is also known that L-lactide and D-lactide form a eutectic crystal. Their structures and melting points are set forth in Figure 13. Lactide can be purified by re-crystallization in a solvent such as toluene [74, 75], by melt crystallization [76] or by distillation [77, 78]. In any of them, first it is aimed to remove the straight-chain lactic acid compounds. The purification methods are different from each other in that they can remove different isomers of lactide. When lactic acid is to be produced, either its L- or D-isomer is intended to be produced in high purity. Actually, however, the product is not always obtained with an optical purity of 100%. Further, when lactide is synthesized, lactic acid may be racemized. Accordingly, the obtained crude lactide comprises,

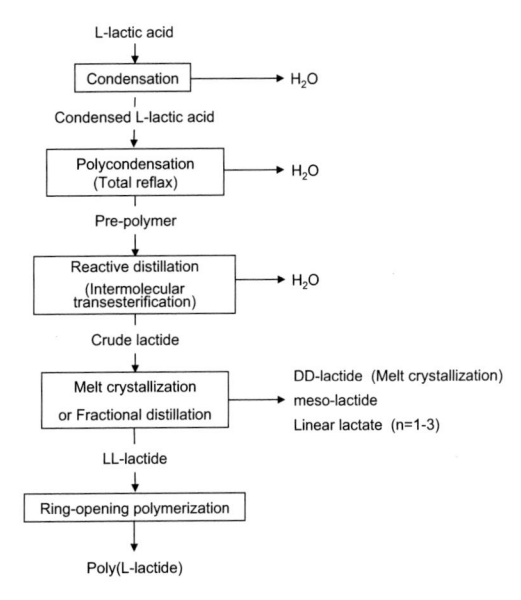

Figure 12 Ring-opening polymerization of purified lactide by melt crystallization or fractional distillation

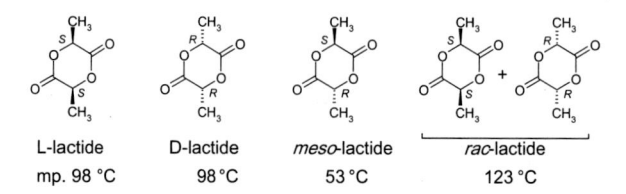

Figure 13 Structure and melting point of lactides

for example, mainly L,L-lactide and a few percent of contaminants such as *meso*-lactide and D,D-lactide. In the melt crystallization, the *meso*-lactide is removed based on the difference of the melting points set forth in Figure 13. Further, the contaminating D,D-lactid in a small amount forms a eutectic crystal with L,L-lactide. Since having a melting point higher than L,L-lactide by 20°C or more, the eutectic crystal can be removed. On the other hand, since D,D-lactide has the same vapor pressure as L,L-lactide, it cannot be separated by distillation. This means that PLLA synthesized from L-lactide purified by crystallization generally has higher crystallinity than that from L-lactide obtained by distillation. Here, it should be noted that the lactide method has problems such as racemization of lactic acid and yield depression in purification of lactide.

(2) Lactide azeotropic method [79] (Figure 14)

In the titled method, first lactic acid is kept condensed in xylene or toluene until straight-chain lactic acid dimer is maximally formed, and then the solvent and lactide are subjected to azeotropic distillation. This method comprises operations at relatively low temperatures, and hence the optical purity of the material lactic acid can be easily maintained. Further, steam can be used as a heat source of the reaction plant, and since the operations can be carried out at room temperature, vacuum equipments are unnecessary. In addition to those merits, it is also advantageous that crystallization for purification can be simultaneously performed in the distilled solvent. On the other hand, however, the rate of conversion into lactide is as low as about 40% and hence it is necessary to circulate unreacted material. Further, it is also disadvantageous to need enough

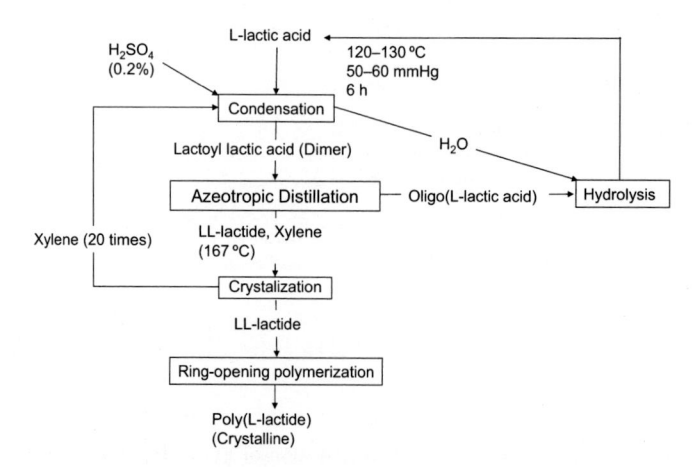

Figure 14 Ring-opening polymerization of purified lactide by azeotropic distillation

energy to heat and evaporate a large amount of the solvent.

If the catalyst remains in the polymer, it may generate lactide again in postforming or may cause coloring. The amount of catalyst is therefore minimized and instead the reaction time is prolonged in industrial production. In the lactide method, there is ring-chain equilibrium between lactide and the polymer and hence some lactide remains after the ring-opening polymerization. When the polymer is subjected to forming such as spinning, the unreacted lactide may cause snap-off of the spun thread. In view of that, after the ring-opening polymerization, lactide is sublimated or washed away with acetone. Since Sn at the growing polymer end generates again lactide in the forming, the amount of catalyst is minimized and instead the reaction time is prolonged in industrial production. In a laboratory, however, a small amount of hydrochloric acid is added to acetone so that Sn is removed in the form of hydrochloride salt. The remaining lactide has a plasticizing effect, and is hydrolyzed by moisture in air into lactic acid or lactic acid dimer to accelerate hydrolysis of the polymer.

3. 1. 4 Direct polycondensation
Solution polymerization (Figure 15)

The direct solution polymerization method has been known since long time ago. In 1955, Kleine applied a patent in which lactic acid was subjected to azeotropic dehydration condensation with toluene or xylene and then the azeotropic solvent was dehydrated with silica-gel and refluxed [80]. Recently, this method has been modified by replacing the solvent with a high-boiling point solvent such as diphenyltoluene [81], but the modified method comprises complicated post-treatments such as pelletization and solvent-washing for removing the high-boiling point solvent [82]. Further, it is also a problem that the resultant polymer has a low molecular weight and poor crystallinity. Since diphenyltoluene is a θ-solvent of poly(lactic acid), poly(lactic acid) is soluble therein at a high temperature but insoluble enough to precipitate at a low temperature. Accordingly, the produced polymer is easily isolated and further lactide in the product is also easily removed.

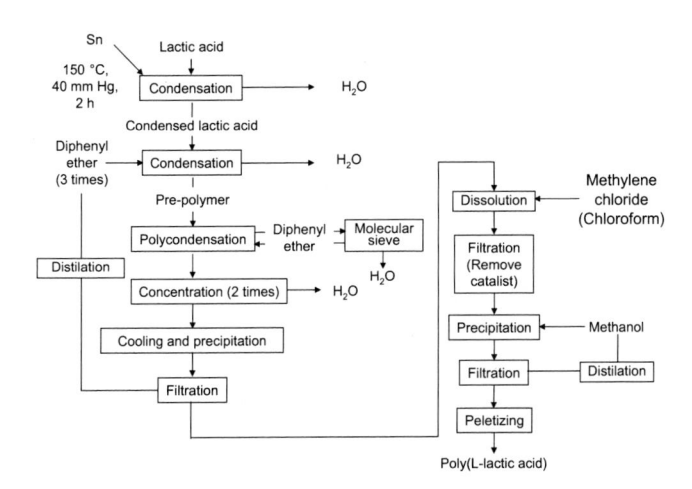

Figure 15 Direct polymerization with solvent

Melt/solid-state polymerization

The polymerization process will be simplified if it is realized that lactic acid is esterfied directly through one-step polycondensation without forming lactide. However, this polycondensation had been thought to be impossible and hence had never been studied for long time. It is not until recently that it has become possible to synthesize PLLA of high molecular weight by direct dehydration polycondensation of lactic acid [83]. This, nevertheless, has not been industrialized yet because of some problems. For example, a large amount of catalyst is used to impair transparency and to lower the glass transition point, and the reaction is carried out at such a high temperature and for such a long time that lactic acid is racemized during the reaction. Further, since lactic acid as the raw material is produced by fermentation, impurities such as sugars and proteins are contained therein and they remain in the resultant polymer to cause coloring and degradation. Meanwhile, another process is developed in which an intermediate polymer having a middle molecular weight is synthesized by melt polymerization and then molded into pellets, which are subsequently subjected to solid-state polymerization [84, 85]. The solid-state polymerization is a technique that has been already industrially applied to manufacture of PET. In this process, a polymer having a pelletizable molecular weight is synthesized by melt polymerization and thereafter the pelletized polymer is further polymerized (Figure 15). Before the polymerization of the pellets, the polymer is crystallized so as to avoid agglutination of the pellets. They applied this melt/solid-state polymerization to production of poly(lactic acid), and thereby succeeded in synthesizing poly(lactic acid) of high molecular weight.

3. 1. 5 Conclusion

Poly(lactic acid) is becoming popular as a bio-based plastic material. In addition, poly(lactic acid) is also known to be used as a bio-medical material and further to be a bio-decomposable and piezoelectric/pyroelectric material. From those viewpoints, various studies are being made at present.

REFERENCES

1. A. Eliasson *et al.*, *Enzyme Microb. Technol.*, **29**, 288 (2001)
2. B. Johansson *et al.*, *Appl. Environ. Microbiol.*, **67**, 4249 (2001)
3. M. Kuyper *et al.*, *FEMS Yeast Res.*, **4**, 655 (2004)
4. M. Kuyper *et al.*, *FEMS Yeast Res.*, **5**, 399 (2005)
5. M. Kuyper *et al.*, *FEMS Yeast Res.*, **5**, 925 (2005)
6. C. Martín *et al.*, *Enzyme Microb. Technol.*, **31**, 274 (2002)
7. M. Sonderegger *et al.*, *Appl. Environ. Microbiol.*, **70**, 2892 (2004)
8. R. Verho *et al.*, *Appl. Environ. Microbiol.*, **69**, 5892 (2003)
9. B. S. Dien *et al.*, *Enzyme Microb. Technol.*, **23**, 366 (1998)
10. B. S. Dien *et al.*, *Appl. Biochem. Biotechnol.*, **84**, 181 (2000)
11. M. A. Cotta and T. R. Whitehead, *Curr. Microbiol.*, **36**, 183 (1998)

12. S. Kumar and D. Ramón, *FEMS Microbiol. Lett.*, **135**, 287 (1996)
13. S. A. Lemmel *et al.*, *Enzyme Microb. Technol.*, **8**, 217 (1986)
14. R. Morosoli *et al. F.*, *FEMS Microbiol. Lett.*, **57**, 57 (1989)
15. A. Peciarová *et al.*, *Biochim. Biophys. Acta*, **716**, 391 (1982)
16. P. L. Pham, P. L. *et al.*, *Ind. Crops Prod.*, **7**, 195 (1998)
17. F. Taguchi *et al.*, *J. Ferment. Bioeng.*, **81**, 178 (1996)
18. S. Virupakshi *et al.*, *Proc. Biochem.*, **40**, 431 (2005)
19. J. K. Yang *et al.*, *J. Mol. Biol.*, **335**, 155 (2004)
20. H. Ohara and T. Yoshida, *Appl. Microbiol. Biotechnol.*, **40**, 258 (1993)
21. T. B. Platt and E. M. Foster, *J. Bacteriol.*, **75**, 453 (1958)
22. T. D. Thomas, *et al.*, *J. Bacteriol.*, **138**, 109 (1979)
23. WHO food additives series No.5 (1974)
24. A. Maanome *et al.*, *J. Gen. Appl. Microbiol.*, **44**, 371 (1998)
25. J. P. De Boer *et al.*, *Appl. Microbiol. Biotechnol.*, **34**, 149 (1990)
26. H. Ohara and M. Yahata, *J. Ferment. Bioeng.*, **81**, 272 (1996)
27. P. M. Yin *et al.*, *J. Ferment. Bioeng.*, **84**, 249 (1997)
28. S. K. Bhattacharyya *et al.*, *Ind. Eng. Chem. Prod. Res. Develop.*, **9**, 92 (1970)
29. S. Dequin and P. Barre, *Bio/Technology*, **12**, 173 (1994)
30. D. Porro *et al.*, *Biotechnology Progress*, **11**, 294 (1995)
31. N. Ishida *et al.*, *Appl. Environ. Microbiol.*, **71**, 1964 (2005)
32. S. Zhou *et al.*, *Appl. Environ. Microbiol.*, **69**, 399 (2003)
33. S. Taguchi *et al.*, *Proc. Natl. Sci. U.S.A.*, **105**, 17323 (2008)
34. M. Yamada *et al.*, *Biomacromolecules*, **10**, 677 (2009)
35. S. Miura *et al.*, *J. Biosci. Bioeng.*, **97**, 19 (2004)
36. E. M. filachione and E.J. Costello, *Ind. Eng. Chem.*, **44**, 2189 (1952)
37. H. Ohara *et al.*, *Seibutsu-kogaku kaishi*, **79**, 142 (2001)
38. W. A. Carothers *et al.*, *J. Am. Chem. Soc.*, **54**, 761 (1932)
39. H. R. Kricheldorf and A. Serra, *Polym. Bull.*, **14**, 497 (1985)
40. J. W. Leenslag and A. J. Pennings, *Macramol. Chem.*, **188**, 1809 (1987)
41. K. Shinno *et al.*, *Macromolecules*, **30**, 6438 (1997)
42. M. Bero *et al.*, *J. Polym. Sci., Part A., Polym. Chem.*, **37**, 4038 (1999)
43. M. Ryner *et al.*, *Macromolecules*, **34**, 3877 (2001)
44. A. J. Nijenhuis *et al.*, *Macromolecules*, **25**, 6419 (1992)
45. G. Schwach *et al.*, *J. Polym. Chem., Part A: Polym Chem.*, **35**, 3431 (1997)
46. H. R. Kricheldorf *et al.*, *Polymer*, **36**, 1253 (1995)
47. F. E. Kohn *et al.*, *J. Appl. Polym. Sci.*, **29**, 2465 (1984)
48. M. B. Bassi *et al.*, *Polym. Bull.*, **24**, 227 (1990)
49. M. Stolt *et al.*, *Macromolecules*, **32**, 6412 (1999)
50. A. Kowalski *et al.*, **33**, 7359 (2000)
51. M. Ryner *et al.*, *Macromolecules*, **34**, 3877 (2001)
52. A. Kowalski *et al.*, *Macromolecules*, **31**, 2114 (1998)
53. J. Baran *et al.*, *Macromol. Symp.*, **123**, 93 (1997)
54. S. Matsumura *et al.*, *Macromol. Symp.* **130**, 285 (1998)
55. S. Matsumura *et al.*, *Macromol. Rapid. Commun.*, **18**, 477 (1997)
56. M. Hans *et al.*, *Macromol. Biosci.*, **9**, 239 (2009)
57. O. Coulembier *et al.*, *Macromolecules*, **39**, 5617 (2006)
58. A. P. Dove *et al.*, *J. Am. Chem. Soc.*, **127**, 13798 (2005)

59. R. C. Pratt *et al.*, *J. Am. Chem. Soc.*, **128**, 4556 (2006)
60. A. P. Dove *et al.*, *Polymer*, **47**, 4018 (2006)
61. A. P. Dove *et. al.*, *Chem Commun.*, 2881-2883, D0I: 10.1039/b601393g (2006)
62. C. Wang *et al.*, *Biomaterials*, **25**, 5797 (2004)
63. H. Li *et al.*, *J. Polym. Sci, Part A: Polym. Chem.*, **42**, 3775 (2004)
64. H. Li *et al.*, *Ind. Eng. Chem. Res.*, **44**, 8641 (2005)
65. N. Spassky *et al.*, *Macromol. Chem. Phys.*, **197**, 2627 (1996)
66. C. P. Radano *et al.*, *J. Am. Chem. Soc.*, **122**, 1552 (2000)
67. T. M. Ovitt *et al.*, *J. Polym. Sci, Part A: Polym. Chem.*, **38**, 4686 (2000)
68. T. M. Ovitt *et al.*, *J. Am. Chem. Soc.*, **124**, 1316 (2002)
69. N. Nomura *et al.*, *J. Am. Chem. Soc.*, **124**, 5938, (2002)
70. P. Hormnirun *et al.*, *J. Am. Chem. Soc.*, **126**, 2688, (2004)
71. Z. Tang *et al.*, *Biomacromolecles*, **5**, 965 (2004)
72. B. M. Chamberlain *et al.*, *J. Am. Chem. Soc.*, **123**, 3229, (2001)
73. C-X. Cai *et al.*, *Chem. Commun.*, **3**, 330 (2004)
74. M. Ogaito, Japanese Patent Laid-Open Publication H7-118259 (1995)
75. S. Sawa *et al.*, Japanese Patent Laid-Open Publication H7-138253 (1995)
76. S. Ohara *et al.*, Japanese Patent Laid-Open Publication H8-208638 (1996)
77. P. R. Gruber *et al.*, USP 5,142,023 (1992)
78. P. R. Gruber, USP 5,359,026 (1993)
79. D. W. Verser *et al.*, USP 5,420,304 (1993)
80. J. Kleine, BP, 779,291 (1955)
81. K. Enomoto *et al.*, Japanese Patent Laid-Open Publication H6-65360 (1994)
82. M. Ajioka *et al.*, *Bull. Chem. Soc. Jpn.*, **68**, 2125 (1995)
83. N. Yanagisawa *et al.*, Japanese Patent Laid-Open Publication H7-828432 (1995)
84. S.-I. Moon *et al.*, *Polymer*, **42**, 5059 (2001)
85. S.-I. Moon *et al.*, *Polym. Commun.*, **42**, 5059 (2001)

3.2 Stereocomplex PLA
Hideki Yamane

3.2.1 Introduction

Poly(L-lactic acid), an aliphatic polyester derived from renewable resources, is a crystalline polyester having its T_m around 180°C. Because of its biocompatibility and biodegradability, PLLA has been utilized for surgical implant materials and drug delivery systems as well as ecological materials. However since the hydrolytic degradation rate of highly oriented PLLA products with a high crystallinity is rather low and they can keep their shape and mechanical properties for fairly long time [1], PLLA is expected to be utilized for general purposes such as textile fibers, films and other products. One of the problems, which may be a major obstacle to the practical applications of PLLA, is its low thermal resistance. The fundamental solution to this problem may be a utilization of the stereocomplex.

Ikada et al.[2] first reported that the 1/1 blend of poly(L-lactic acid) (PLLA) and poly(D-lactic acid) (PDLA), which is an enantiomer of PLLA, produces a stereocomplex with T_m around 230°C. While pure PLLA and PDLA crystallize in an orthorhombic form with a 10/3 helix in their conformation [3], the stereocomplex has a triclinic form with 3/1 helix [4,5]. Since this first report, they intensively studied the effects of various parameters including blending ratio [6–13], molecular weight [6–9], optical purity [10,11], and blending condition [6,8–10] on the stereocomplexation between PLLA and PDLA. Although many papers reported that the 1/1 blend of PLLA and PDLA shows only a melting of the stereocomplex in the thermal analyses, this is not always the case especially when PLLA and PDLA with high molecular weights are melt blended.

Takasaki et al.[14] reported that the high speed melt-spun fibers of PLLA/PDLA blend contain both homo-crystal of PLLA or PDLA and the stereocomplex, and the annealing gave a fiber mainly consisting of highly oriented stereocomplex crystal. Furuhashi et al.[15] also showed a similar result for the melt-spun PLLA/PDLA blend fibers and demonstrated that the homo-crystal melts and recrystallizes into stereocomplex when the blend fibers were annealed at an elevated temperature. One of the present authors pointed out that although the homo-crystal phase presented in the PLLA/PDLA blend fiber can be transformed into the stereocomplex, the degree of the crystalline orientation was significantly reduced resulting in the deterioration of the mechanical property [16]. Zhang et al.[17] succeeded in preparing the PLLA/PDLA blend films with a highly oriented stereocomplex phase by drawing and annealing at optimum conditions and demonstrated the phase transition phenomenon from α- or α'-form to the stereocomplex. However they also reported that the blend comprising PLLA and PDLA with molecular weights higher than 1×10^5, which generally utilized for the industrial melt-spinning process, tends to have the co-existence of both homo- and stereocomplex crystal phases.

A special melt blending process was recently developed to obtain PLLA/PDLA melt blend, in which the stereocomplex phase consistently forms without forming any trace of PLLA or PDLA homo-crystal phase [18]. Further the PLLA/PDLA blend obtained was melt-processed into fibers and films. Changes in the structure and the property of the products during various processes such as drawing and annealing were investigated [18,19].

3. 2. 2 Melt-blending of PLLA and PDLA

Masaki *et al.* have prepared the PLLA/PDLA blend which only shows SC in a repetitive heating and cooling processes by kneading equal amounts of PLLA and PDLA at a temperature between the melting temperatures of the homo-crystal and the stereocomplex and succeedingly melt mixing at a temperature higher than the melting temperature of the stereocomplex [18].

Melt blending of PLLA and PDLA were carried out using a melt mixer as schematically shown in Figure 1. When the melt blending was carried out at temperatures between 170 and 200°C, granular solids were extruded from the mixer. On the other hand, the blending at higher temperatures between 210 and 240°C gave molten extrudates. Typical DSC heating curves of these extrudates are shown in Figure 2. The granular solids prepared at lower temperatures were

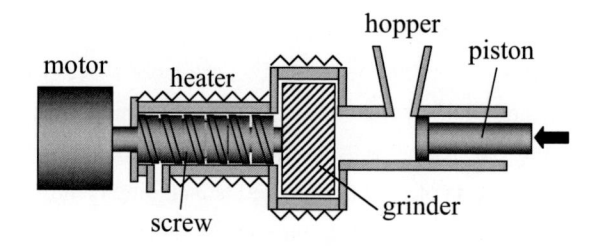

Figure 1   Schematic diagram of the melt mixer used in this study

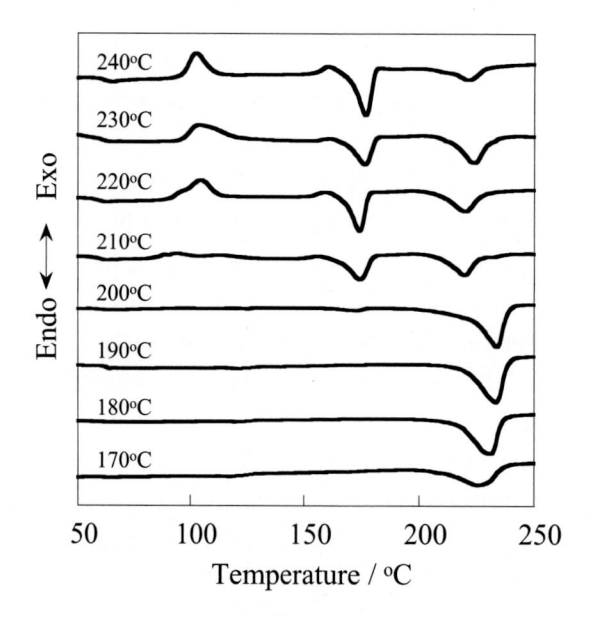

Figure 2   DSC curves of the blends prepared at various temperatures

highly crystalline and showed a large endothermic peak at 230°C which can be attributable to the meting of the stereocomplex without indicating any crystallization exotherm. The glassy solids quenched from the melts mixed at temperatures ranging from 210 to 240°C showed a large crystallization peak at 100°C and two endothermic peaks at 175 and 220°C.

WAXD spectra of the granular blends prepared at various temperatures are shown in Figure 3. The PLLA showed reflections at 2θ around 12°, 17°, 19°, and 22.5° indexed as (101), (110)/(200), (201)/(111), and (102)/(210) of the α-form homo-crystal. On the other hand, the granular solids showed reflections at 2θ around 12°, 20°, and 23° indexed as (100)/(010)/(-110), (110)/(-120)/(-210), and (200)/(020)/(-220) of the stereocomplex crystal, respectively and no reflection from the homo-crystal was detected. These results indicate the effective formation of the stereocomplex in the blends prepared at temperatures between the T_m of PLLA homo-crystal and that of the stereocomplex.

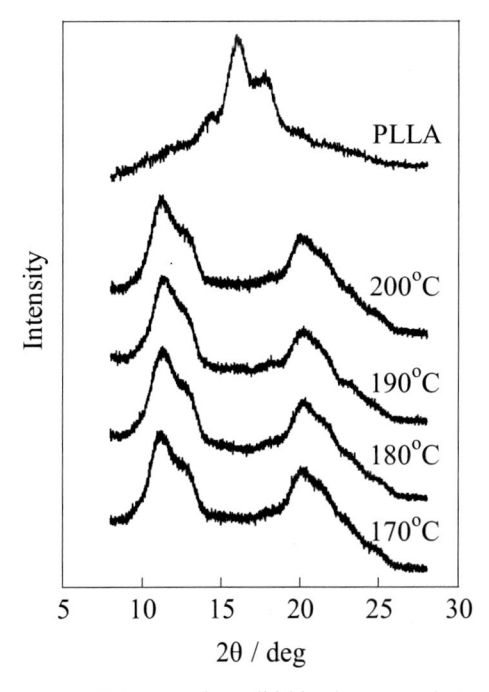

Figure 3 WAXD spectra of the granular solid blends prepared at various temperatures

The granular blends prepared have to be processed into various products at an elevated temperature at which the stereocomplex melts. The DSC curves of the blend extruded at 250°C and quenched are shown in Figure 4(a). The blend prepared at 190°C in advance showed a crystallization peak at 100°C and only a melting endotherm at 220°C. Even in the following cooling process from the melt at a high cooling rate, it showed a strong crystallization peak at 120°C. Considering that the pure PLLA cooled from the melt does not show a clear crystallization peak even at a much lower cooling rate, this blend has a quite high crystallization rate. It again showed only a melting endotherm of the stereocomplex in the 2nd heating process without indicating any melting endotherm of the homo-crystal. These results indicate that the kneading of the granular solid blend at an elevated temperature gave a blend which produces only the stereocomplex consistently even in the repetitive melting-solidifying processes. On the other hand, the blend prepared at 240°C in advance showed the melting peaks of both the homo- and the stereocomplex as was generally observed in the melt blend of PLLA and PDLA as shown in Figure 4(b).

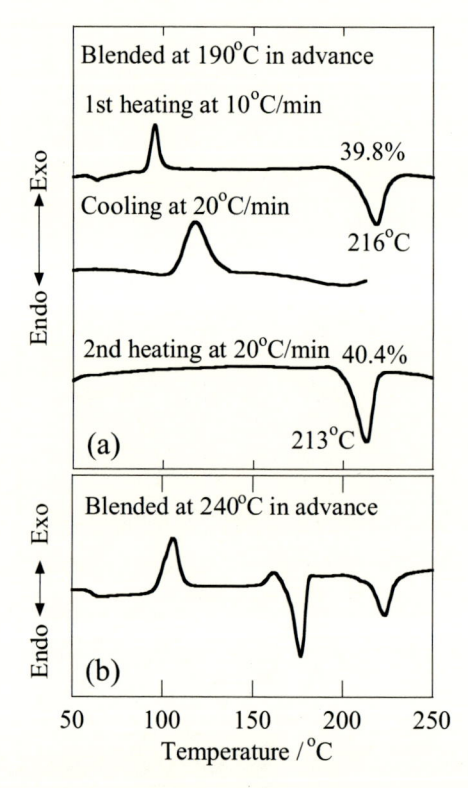

Figure 4   DSC curves of the blend kneaded at 250°C. (a) Blend prepared at 190°C and (b) blend prepared at 250°C in advance

The 1st blending process at a low temperature achieved the mixing of PLLA and PDLA in a molecular level and this state was frozen in the stereocomplex crystal phase. The 2nd blending process at a high temperature stabilized this state due to the formation of the block copolymer through the trans-esterification occurred at the interface between PLLA and PDLA.

3. 2. 3 Melt-spinning of PLLA/PDLA blend

The granular solid blend obtained at 190°C was melt-spun into fibers at 250°C. As-spun fiber quenched was in an amorphous state. Drawing was carried out at various temperatures ranging from 60°C to 100°C. Most drawn fibers kept their amorphous state except for those drawn × 3 at 60°C and × 4 at 80°C as shown in Figure 5. These two fibers showed an oriented broad crystalline reflection centered at 2θ = 17°. This may be a characteristic of the disordered α'-form crystal. Zhang et al.[17] reported that the drawing of the PLLA/PDLA blend film produced the disordered α'-form crystal phase.

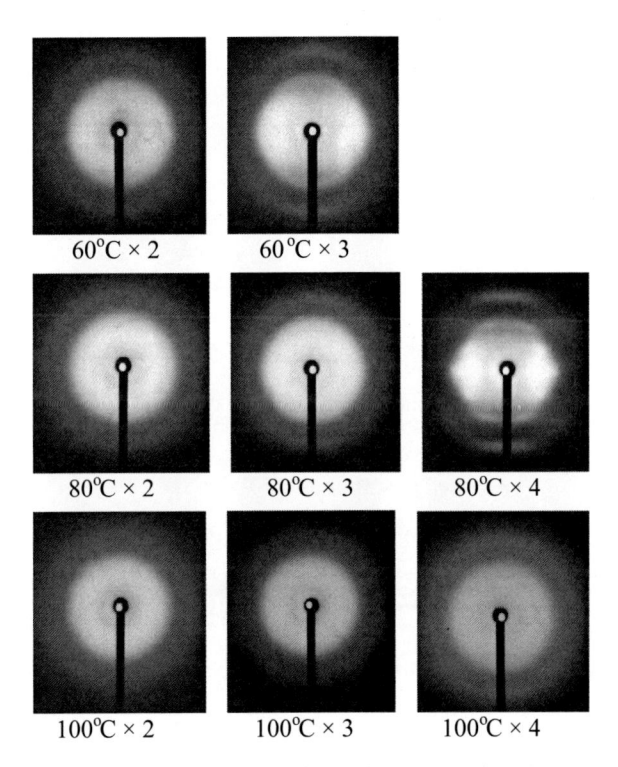

Figure 5 WAXD pattern of the fibers drawn at various temperatures

The fibers drawn at various conditions were annealed at 140, 160 and 180°C for 5 sec under tension. Most of the amorphous drawn fibers immediately melted at these temperatures. However the drawn fibers which showed a broad crystalline reflection were successfully annealed. It is striking to note that the crystalline form was completely transformed from the α'-form to the highly oriented stereocomplex as seen in Figure 6. The broad peak detected for the drawn fiber at 2θ = 17° completely disappeared and the new peaks appeared in the equatorial spectra at 12, 20 and 23°. These peaks tended to be much sharper with increasing annealing temperature although the change in the crystalline orientation was not determined clearly from these WAXD patterns.

This sort of change in the structure reflected the mechanical property. The strength and the modulus shown in Figure 7(a) and (b) increased and the elongation at break (not shown) slightly decreased with increasing annealing temperature.

One of the most important objectives was to obtain the PLA fiber with high thermal resistance. Figure 8 compares the dynamic viscoelasticity of the stereocomplex and PLLA fibers. The stereocomplex fiber was prepared by drawing PLLA/PDLA blend fiber at 80°C to × 4 and annealed at 180°C. PLLA fiber was drawn at 80°C to × 4 and annealed at 120°C. While PLLA fiber showed a significant decrease in the storage modulus around its T_g and melted down at around 160°C, the stereocomplex fiber kept its modulus fairly high up to 185°C.

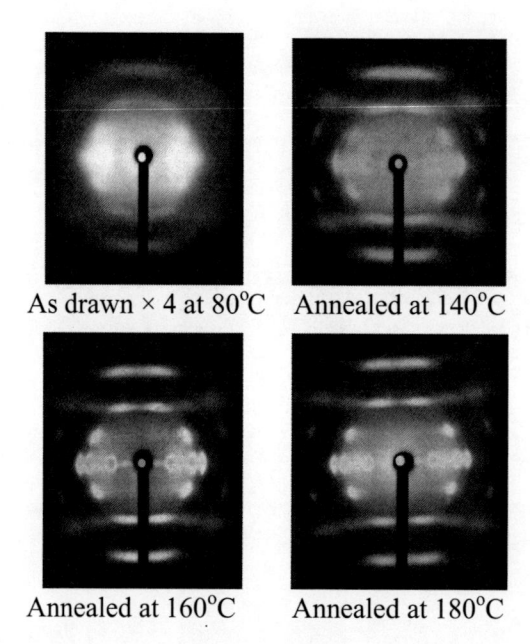

Figure 6 WAXD patterns of the fibers drawn and annealed at various temperatures

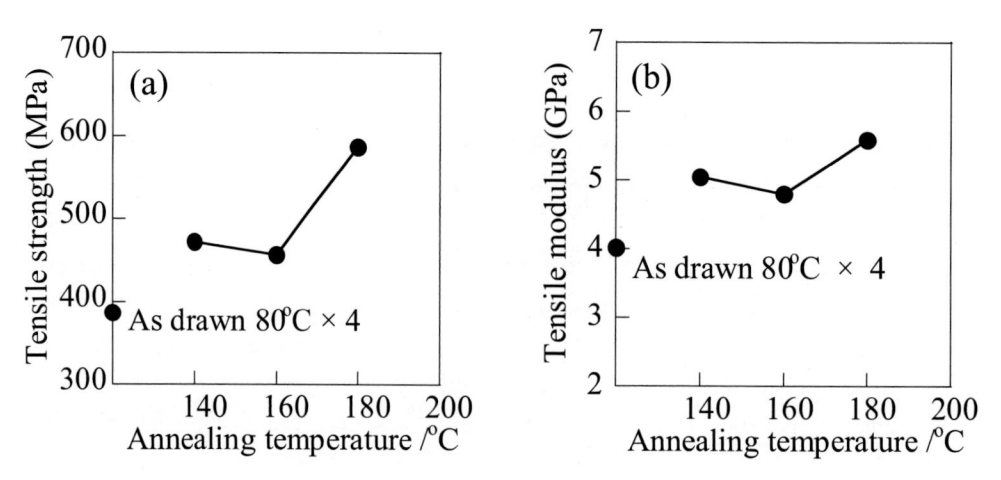

Figure 7 Tensile strength (a) and tensile modulus (b) of the drawn and annealed fibers

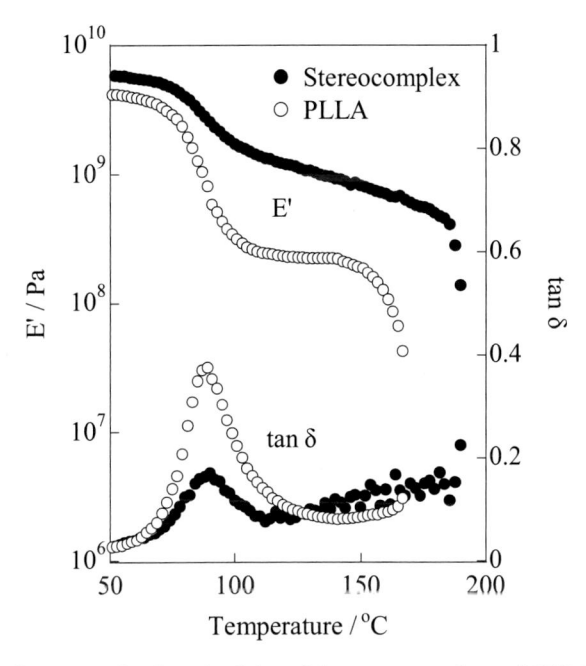

Figure 8 Dynamic viscoelasticity of the stereocomplex and PLLA fibers

3. 2. 4 Biaxially oriented PLLA/PDLA blend films

Equal amount of PLLA and PDLA were melt blended in the procedure described in the previous section. Compression molded PLLA/PDLA blend films were biaxially stretched by using a manual biaxial drawing apparatus shown in Figure 9. Although the films can be biaxially drawn at temperatures between 60°C to 80°C, it was most easily and uniformly drawn to 2.5 × 2.5 at 70°C. Drawing at temperatures higher than 90°C was not possible due to the rapid crystallization of the stereocomplex. Figure 10(a)~(c) show the mechanical property of the films biaxially drawn and annealed. Undrawn film was rather brittle and showed a low tensile strength and a low elongation to break. The tensile strength was increased with increasing draw ratio, although the

Figure 9 Manual biaxial drawing apparatus

tensile modulus did not change significantly. The films drawn at 70°C were successfully annealed at temperatures 180°C and 190°C for 10 – 60 sec, although it was not uniformly deformed during the annealing process at 200°C.

Figure 11(a) and (b) show the DSC curves of the blend films undrawn and drawn to 2.5 × 2.5 annealed at 190°C for various periods of time. Both undrawn and drawn films showed a crystallization peak around 90°C and a melting peak around 215°C before annealing. The crystallization peak of the undrawn film did not disappear even after annealing for 60 sec. However, that of the drawn film completely disappeared after annealing for 30 sec indicating complete formation of SC at 190°C within 30 sec. WAXD patterns shown in Figure 12 also indicate that almost amorphous state of the unannealed film transformed to SC during annealing at 190°C within 30 sec. The tensile strength of drawn film increased with annealing time up to 30 sec and the tensile modulus increased with annealing time up to 10 sec. The longer annealing time at an elevated temperature deteriorated the mechanical property.

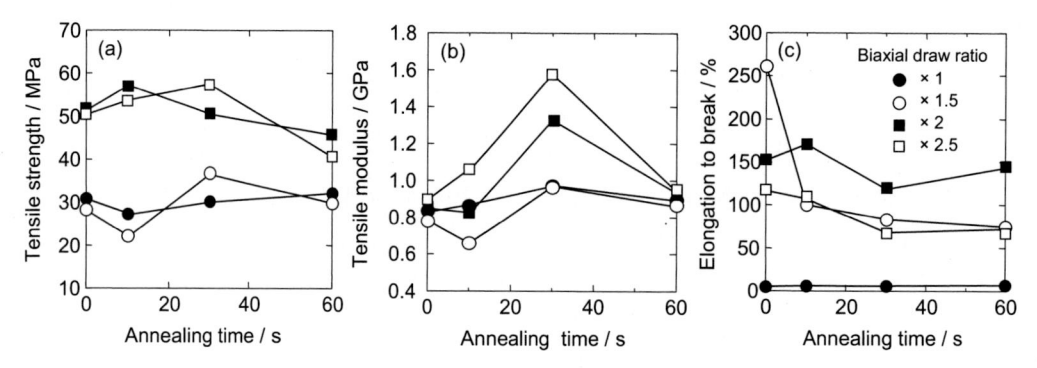

Figure 10 Mechanical property of the PLLA/PDLA blend films drawn and annealed for various periods of time. (a) Tensile strength, (b) tensile modulus, and (c) elongation to break

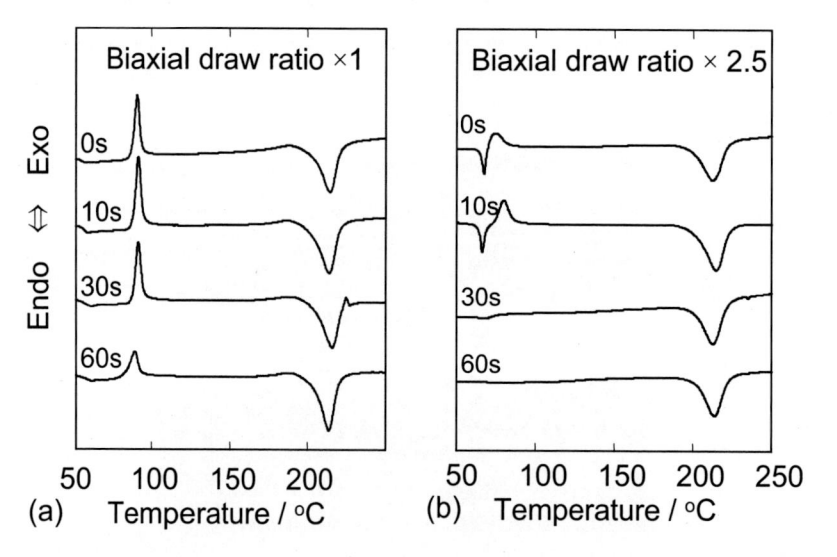

Figure 11 DSC curves of undrawn and drawn PLLA/PDLA blend films annealed for various periods of time

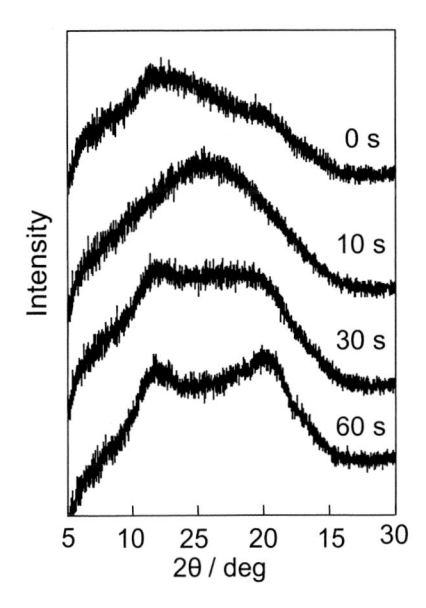

Figure 12 WAXD patterns of PLLA/PDLA blend films drawn to 2.5 × 2.5 and annealed at 190°C for various periods of time

3. 2. 5 Conclusion

High molecular weight poly(L-lactic acid) and poly(D-lactic acid) were melt mixed in a special procedure into a blend which produces the stereocomplex consistently without indicating any trace of the occurrence of the homo-crystal. The granular blend with high stereocomplex content was obtained by mixing PLLA and PDLA at a temperature between T_{ms} of the homo-crystal and the stereocomplex. In this blending process, the mixing of PLLA and PDLA in a molecular level was achieved and this state was frozen. Kneading this granular blend again at high temperature gave a PLLA/PDLA blend which forms the stereocomplex consistently even in the repeated melting-solidifying processes.

Melt-spun fibers and the compression molded films of the blend showed an α'-form crystal which immediately transformed to the stereocomplex in the annealing process at elevated temperatures without relaxing the molecular orientation resulting in the superior thermal and mechanical properties.

References

1. Y. Doi, *ed.*, "Biodegradable Plastic Handbook", NTS Inc., (1995)
2. Y. Ikada, K. Jamshidi, H. Tsuji, S. –H. Hyon, *Macromolecules*, **20**, 904 (1987)
3. P. DeSantis, A. J. Kovacs, *Biopolymers*, **6**, 209 (1968)
4. T. Okihara, M. Tsuji, A. Kawaguchi, K. Katayama, H. Tsuji, S. –H. Hyon, Y. Ikada, *J. Macromol. Sci.,-Phys.*, **B30**, 119 (1991)
5. T. Okihara, A. Kawaguchi, H. Tsuji, S. –H. Hyon, Y. Ikada, K. Katayama, *Bull. Inst. Chem. Res.*,

Kyoto Univ., **66**, 271 (1988)
6. H. Tsuji, F. Horii, S. –H. Hyon, Y. Ikada, *Macromolecules*, **24**, 2719 (1991)
7. H. Tsuji, S. –H. Hyon, Y. Ikada, *Macromolecules*, **24**, 5651 (1991)
8. H. Tsuji, S. –H. Hyon, Y. Ikada, *Macromolecules*, **25**, 2940 (1992)
9. H. Tsuji, Y. Ikada, *Macromolecules*, **26**, 6918 (1993)
10. S. Brouchu, R. E. Prud'homme, I. Barakat I, R. Jerome, *Macromolecules*, **28**, 5230 (1995)
11. H. Tsuji, Y. Ikada, *Macromol. Chem. Phys.*, **197**, 3483 (1996)
12. S. C. Schmidt, M. A. Hillmyer, *J. Polym. Sci., Part B, Polym. Phys.*, **39**, 300 (2001)
13. H. Yamane, K. Sasai, *Polymer*, **44**, 2569 (2003)
14. M. Takasaki, H. Ito, T. Kikutani, *J. Macromol. Sci. Part B-Phys.*, **B42**, 403-420 (2003)
15. Y. Furuhashi, Y. Kimura, N. Yoshie, H. Yamane, *Polymer*, **45**, 5972 (2006)
16. H. Yamane, *Nihon Reoroji Gakkaishi*, **27**, 4, 213 (1999)
17. J. Zhang, K. Tahiro, H. Tsuji, A. J. Domb, *Macromolecules*, **40**, 1049 (2007)
18. D. Masaki, Y. Fukui, K. Toyohara, M. Ikegame, B. Nagasaka, and H. Yamane, *Sen'i Gakkaishi*, **64**, 8, 212-219 (2008)
19. Y. Fukui, D. Masaki, J. –J. Lee, J. –C. Lee, H. Yamane, *Sen'i Gakkaishi*, submitted.

3.3 Polyhydroxyalkanoate

Tadahisa Iwata, Takeharu Tsuge, Sei-ichi Taguchi, Hideki Abe and Toshihisa Tanaka

3.3.1 General introduction

A wide variety of bacteria synthesize an optically active polymer of (R)-3-hydroxybutyric acid and accumulate it as intracellular carbon and energy storage material. Poly[(R)-3-hydroxybutyrate] (P(3HB)) isolated from bacteria is extensively studied as a biodegradable and biocompatible thermoplastic with a melting temperature at ~180°C (Figure 1) [1–4]. P(3HB) was discovered by Maurice Lemoignei of Institute Pasteur [5], France, in the 1920s, and until now many researchers have investigated its physical properties, crystal structure, biosynthesis mechanism, biodegradation mechanism, etc [6]. P(3HB) was initially produced by Imperial Chemical Industries by using an industrial-scale fermentation process in the 1980s. In 1995, the process and related patents were bought by Monsanto and subsequently acquired by Metabolix. Recently, Lenz and Marchessault reported a historical review for the chemical, biochemical and microbial highlights of P(3HB) covering a discovery timespan of 75 years [7].

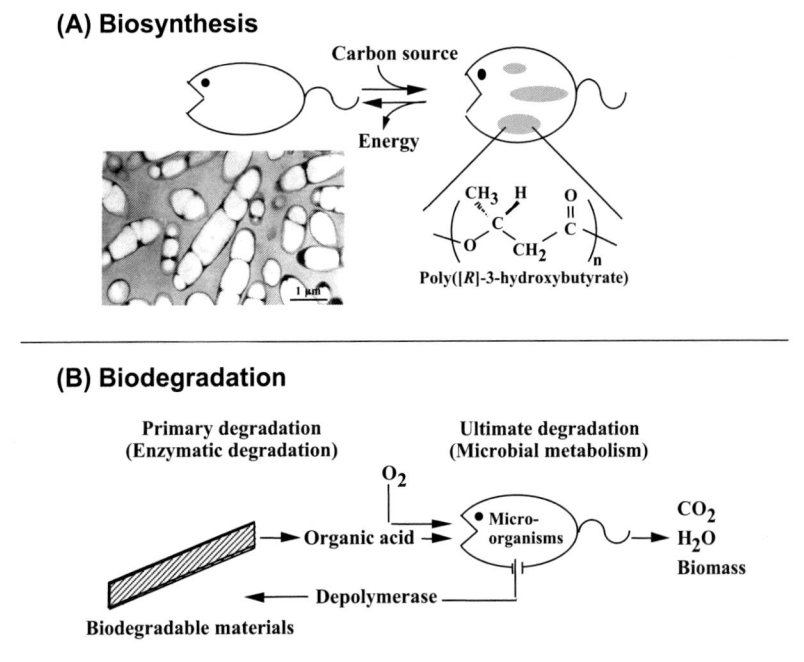

Figure 1 Biosynthesis and biodegradation of microbial polyesters. In set: TEM picture of microorganism accumulating 81% PHA.

3.3.2 Biosynthesis of P(3HB) and its copolymers

Three main PHA biosynthesis pathways identified so far are shown schematically in Figure 2 [8, 9]. The most common pathway I, in which (R)-3HB monomer is generated from a starting substance acetyl-CoA, has been found in a wide range of bacteria accumulating short chain length (scl)-PHA. This pathway has been extensively investigated for *Ralstonia eutropha* [10, 11]. Two acetyl-CoA molecules are condensed to yield acetoacetyl-CoA by 3-ketoacyl-CoA thiolase (PhaA, acetyl-CoA:acetyl-CoA-acetyl transferase: EC 2.3.1.9). The resultant acetoacetyl-CoA is subsequently reduced to (R)-3HB-CoA by an NADH- or NADPH-dependent acetoacetyl-CoA reductase (PhaB, (R)-3HA-CoA dehydrogenase: EC 1.1.1.36). Only (R)-isomers are incorporated as the substrates into the 3HB homopolymer (P(3HB)) catalyzed by the polymerizing enzyme, PHA synthase (PhaC).

Pathway II and III found in various fluorescent pseudomonads, supply mainly medium chain length (mcl)-(R)-3HA monomers from fatty acid β-oxidation and fatty acid biosynthesis intermediates, respectively. The intermediates in these pathways are effectively converted by some specialized enzymes to generate (R)-3HA-CoA monomers for polymerization. As shown in Figure 2, (R)-specific enoyl-CoA hydratase (PhaJ) [12] and (R)-3-HA-ACP-CoA transferase (PhaG) [13] function as metabolic suppliers for (R)-3HA-CoA from trans-2-enoyl-CoA and (R)-3HA-ACP,

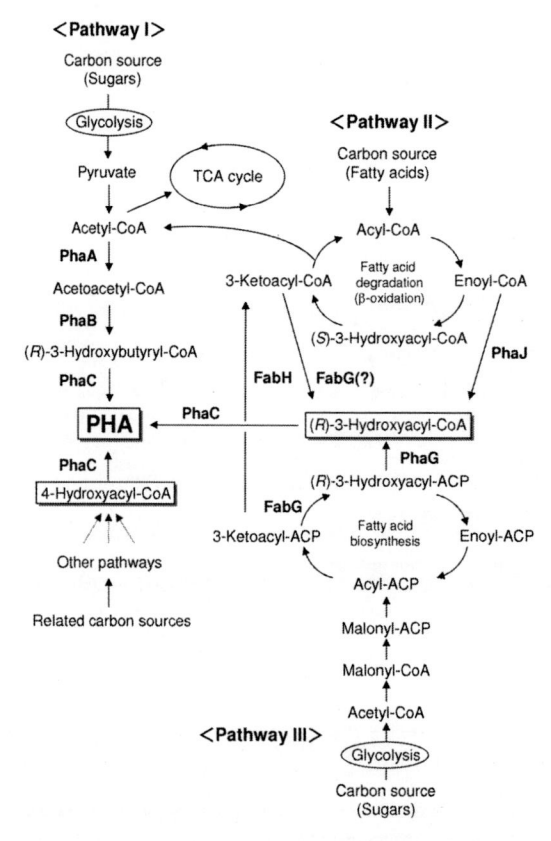

Figure 2 Metabolic pathways that supply various hydroxyalkanoate monomers for PHA biosynthesis.

respectively.

It has also been demonstrated that 3-ketoacyl-ACP reductase FabG can accept not only acyl-ACP but also acyl-CoA as a substrate and is capable of supplying mcl-(R)-3HA-CoA from fatty acid β-oxidation in *Escherichia coli* [14]. FabG, which is known as a homolog of PhaB, has been reported to serve as a monomer supplier for PHA biosynthesis in recombinant *E. coli* [14, 15]. 3-Ketoacyl-acyl carrier protein (ACP) synthase III (FabH) is also a constituent of fatty acid biosynthesis, but not a specialized enzyme for PHA synthesis. It was previously shown that coexpression of the PHA synthase gene and the 3-ketoacyl-ACP synthase III gene (fabH) from *E. coli* led to the production of P(3HB) homopolymer in recombinant *E. coli* grown in the presence of glucose [16].

3. 3. 3 Fermentative production and mechanical properties of PHB and its copolymers

It has been well known that mechanical properties of P(3HB) homopolymer films markedly deteriorate by a process of secondary crystallization [17, 18]. Accordingly, microbial P(3HB) homopolymer has been regarded as a polymer required to copolymerize with other monomer components from the viewpoint of industrial applications because of its stiffness and brittleness [19–21]. As alternative way to improve the physical properties of P(3HB), the incorporation of different hydroxyalkanate (HA) units into the P(3HB) sequence to form PHA copolymers is effective. Various bacteria are capable of synthesizing random copolymers of (R)-3HB with other HA units of C3 to C12, depending on both their intrinsic PHA biosynthesis pathways and the carbon sources used. Figure 3 and Table 1 show the structures and properties of typical PHA copolymers containing the (R)-3HB unit as a constituent, together with P(3HB). A copolymer of (R)-3HB and (R)-3-hydroxyvalerate [(R)-3HV], P(3HB-co-3HV), has been investigated most extensively among the PHA copolymers and already applied to commercial products, because this copolymer is easily produced by feeding propionic acid as a co-substrate to *R. eutropha* [22]. By introducing 20 mol% (R)-3HV units, the melting temperature and glass-transition temperature of the P(3HB-co-3HV) copolymer decrease to from

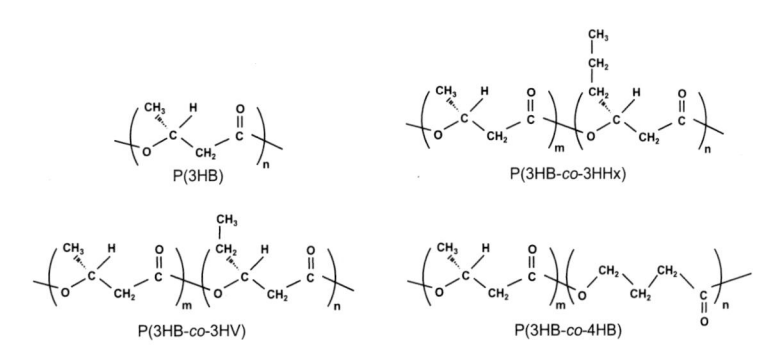

Figure 3 Chemical structure of P(3HB) and its copolymers: P(3HB); poly[(R)-3-hydroxybutyrate], P(3HB-co-3HV); poly[(R)-3-hydroxybutyrate-co-(R)-3-hydroxyvalerate], P(3HB-co-3HHx); poly[(R)-3-hydroxybutyrate-co-(R)-3-hydroxyhexanoate], and P(3HB-co-4HB); poly[(R)-3-hydroxybutyrate-co-4-hydroxybutyrate].

Table 1 Mechanical and thermal properties of PHA films.

Polymer	T_m^a (°C)	T_g^b (°C)	Tensile strength (MPa)	Elongation at break (%)
P(3HB)	177	4	43	5
P(3HB-co-8 mol% 3HV)	165	1	19	35
P(3HB-co-20 mol% 3HV)	145	−1	20	50
P(3HB-co-16 mol% 4HB)	150	−7	26	444
P(3HB-co-64 mol% 4HB)	50	−35	17	590
P(3HB-co-90 mol% 4HB)	50	−42	65	1,080
P(3HB-co-5 mol% 3HHx)	160	−2	32	260
P(3HB-co-10 mol% 3HHx)	127	−1	21	400
P(3HB-co-6 mol% 3HA)^c	133	−8	17	680
Polypropylene (PP)	176	−10	38	400
Low-density polyethylene (LDPE)	130	−36	10	620

^{a} Melting temperature
^{b} Glass-transition temperature
^{c} 3HA units: 3-hydroxydecanoate (3 mol%) and 3-hydroxydodecanoate (3 mol%)

177 to 145°C and 4 to −1°C, respectively. In contrast, its elongation at break increases from 5% to 50%, indicating that the P(3HB-co-3HV) copolymer is more flexible than the P(3HB). Also, the incorporation of 4-hydroxybutyrate (4HB) units into the P(3HB) sequence is more effective in improving the material properties of PHA. P(3HB-co-16 mol% 4HB) copolymer exhibits elongation at break as high as 444% [23]. In the case of (R)-3HB-based copolymers containing (R)-3-hydroxyhexanoate [(R)-3HHx] and longer (R)-3HA (C8-C12), a small amount of the second monomer unit is sufficient to increase their flexibility. It is of interest to note that the P(3HB-co-6 mol% 3HA) copolymer shows similar mechanical properties to low-density polyethylene (LDPE) [24]. Thus, this P(3HB-co-3HA) copolymer is expected to have various commercial applications.

As for P(3HB-co-3HHx), efficient production has been achieved by using recombinant bacterium from vegetable oil as carbon sources. Vegetable oils are desirable feedstock for PHA production by bacteria because of their low material costs (ca. 30 US cents per kg) and relatively high yields of PHA production (0.7–0.8 g-PHA per g-vegetable oil; this value is two-fold higher than that from glucose) [22]. The bacterium *Aeromonas caviae* isolated from soil is capable of producing P(3HB-co-3HHx) copolymer from vegetable oils [25]. However, this bacterium has a poor capability for PHA accumulation (less than 30 wt% of the dry cells). The well-known PHA producing bacterium *R. eutropha* is capable of accumulating PHA at high levels exceeding 80 wt% of the dry cells, but it produces only P(3HB) homopolymer from vegetable oils. Hence, a genetically engineered recombinant *R. eutropha* PHB$^-$4 (PHA negative mutant strain) harboring *A. caviae* PHA synthase gene (*phaC*$_{Ac}$) has been generated [26, 27].

To produce P(3HB-co-3HHx) from vegetable oil (soybean oil) by this recombinant, larger scale fermentation using a 10 L-jar fermentor was carried out [28]. The soybean oil concentration in fermentor was kept at 20 g/L and a phosphorous-limitation for cell growth was used to initiate an efficient PHA accumulation in cells after exponential growth phase. It was succeeded in achieving efficient production of P(3HB-co-5 mol% 3HHx) with high cell concentration of 138 g/L and high PHA content of 74 wt%. The PHA yield was as high as 0.72 g-PHA per g-soybean oil, which is two-fold higher than that from glucose [28].

Based on the above fermentation performances, an environmental life cycle inventory of PHA production from glucose or soybean oil was compared in terms of total fossil energy consumptions and net CO_2 emissions [29]. It has been concluded that the net CO_2 emissions of PHA production from soybean oil (less than 1.0 kg-CO_2/kg-PHA) are relatively lower than those for production of petrochemical plastics such as polyethylene, polypropylene, polystyrene and polyethyleneterephthalate (1.7–3.1 kg-CO_2/kg-plastic). In addition, high yield production of PHA from soybean oil was verified to be cost competitiveness (less than 4 US$/kg-PHA).

3. 3. 4   Ultra-high-molecular-weight P(3HB)

P(3HB), containing repeating units of (R)-3HB, is the most common biological polyester produced by many of bacteria (Figure 1A). Some bacteria represented by *R. eutropha* (currently *Cupriavidus necator*) are able to accumulate P(3HB) up to 80% of the cell dry weight from various carbon sources such as sugars, organic acids, plant oils, and carbon dioxide [22]. The weight-average molecular weight (M_w) of P(3HB) produced by native PHA-producing bacteria is usually in the range of $5–10 \times 10^5$. In some cases, it is known that bacteria are capable of synthesizing very high molecular weight P(3HB) with M_w exceeding 3×10^6 g/mol. This type of P(3HB) is particularly termed ultra-high-molecular-weight P(3HB) [UHMW-P(3HB)] [30, 31].

In general, P(3HB) is a polymer with poor material properties for practical application; it is a highly crystalline and stiff material, thus, it is brittle and has poor elastic qualities. In addition, fibers and films forming from usual P(3HB) are difficult. However, the processability can be improved by increasing molecular weight. UHMW-P(3HB) can be formed strong fibers and films by applying a proper process, thus, it is recognized as a commercially useful PHA material (see below for further details).

The well-established method to produce UHMW-P(3HB) is by *E. coli*. However, P(3HB) synthesis is not an inherent ability for *E. coli*, thus it needs the help of genetic engineering. Expression of the three genes involved in P(3HB) synthesis, 3-ketothiolase gene (*phaA*), NADPH-dependent acetoacetyl-CoA reductase gene (*phaB*) and PHA synthase gene (*phaC*), confers the ability to accumulate P(3HB). *E. coli* has advantages to use as a host for UHMW-P(3HB) production, that is, it possesses a rich genetic background, displays good cell growth and P(3HB) accumulation, and unlike native PHA producing strains, does not mobilize PHA within the cells.

Other than very high molecular weight of P(3HB), several unusual phenomena have been observed for *E. coli* during accumulating P(3HB). One noteworthy change is the shift in the cell morphology to form filaments, probably due to inactivation of an essential cell division protein FtsZ by the presence of P(3HB) granules in the cells. This is regarded as a major obstacle for P(3HB) production by high cell density culture of *E. coli*, because filamentous cells show low growth activity. To overcome this, overproduction of FtsZ protein in the P(3HB)-accumulating cells has been proposed [32]. In fact, this strategy successfully led to cell growth without cell elongation even when recombinants accumulated large amounts of P(3HB). P(3HB) concentration as high as 157 g/L has been achieved in high cell density culture of *E. coli* harbouring *R. eutropha* PHA biosynthesis genes (*phaCAB*$_{Re}$) and additional *ftsZ* gene from glucose as a sole carbon source [32].

As another way to produce UHMW-P(3HB), two-step cultivation consisting of a cell growth phase and a P(3HB) accumulation phase has been proposed. This strategy does not rely on overproduction of FtsZ protein, but controls the expression of P(3HB) synthesis genes (*phaCAB*$_{Re}$) by an inducible gene promoter with the use of a chemical inducer. In cell growth phase, the cells grow in the absence of any stress to obtain P(3HB)-free cell mass as high as possible. In succeeding P(3HB) accumulation phase, P(3HB) synthesis is induced by addition of chemical inducer and P(3HB) accumulation suppresses the cell growth. This strategy does not prevent cell filamentation but allows us to obtain very high yield of UHMW-P(3HB). The concentration and M_w of P(3HB) were 146 g/L and 4.8×10^6, respectively (Figures 4 and 5) [33,34].

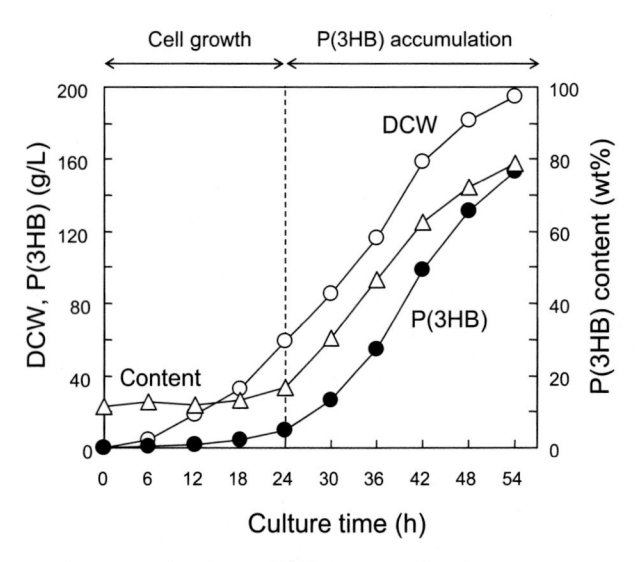

Figure 4 UHMW-P(3HB) production by recombinant *E. coli* using glucose as a carbon source [33]. P(3HB) biosynthesis genes were expressed by addition of chemical inducer at 24 h of cultivation. Molecular weight distribution of P(3HB) is shown in Figure 5. Open circle: dry cell weight (DCW), closed circle: P(3HB) concentration, and open triangle: P(3HB) content in the cells.

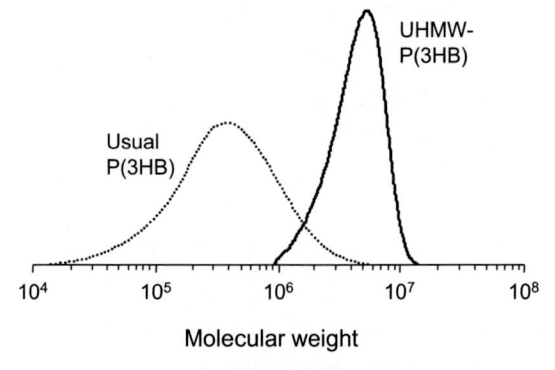

Figure 5 Molecular weight distribution of P(3HB) synthesized by recombinant *E. coli* [UHMW-P(3HB), $M_w=4.8 \times 10^6$, $M_w/M_n=1.5$] [33] and wild-type *R. eutropha* [usual P(3HB)].

It has been demonstrated that the use of low copy number plasmid for expression of P(3HB) biosynthesis genes in *E. coli* is effective in increasing both P(3HB) productivity and its molecular weight rather than the use of high copy number plasmid [34]. This effect may be attributed to production level of PHA synthase in the cells. Additionally, specific culture conditions such as slightly acidic pH and higher temperature result in further increase in M_w up to 20×10^6 [30,35]. PHA synthases that can synthesize UHMW-P(3HB) in *E. coli* are those from not only *R. eutropha* but also *Delftia acidovorans* and *Allochromatium vinosum* [36]. These synthases prefer 3HB unit for polymerization rather than 3HA longer than C5.

3. 3. 5 Structure of P(3HB)

Wide variety of bacteria synthesize an optically active polymer of (*R*)-3-hydroxybutyric acid and accumulate it as intracellular carbon and energy storage material. P(3HB) isolated from bacteria is extensively studied as a biodegradable and biocompatible thermoplastic with a melting temperature at ~180°C. Main crystal structure of P(3HB) has been already reported: crystal system is an orthorhombic, space group is $P2_12_12_1$, and the unit cell is a=0.576 nm, b=1.320 nm and c(fiber axis)=0.596 nm (Figure 6). Molecular chain has a 2/1 helix conformation along the molecular axis (α-form) [37,38].

3. 3. 6 Fibers of P(3HB) and its copolymer

Some research groups have attempted to improve the mechanical properties of P(3HB) films [31,39–44] and fibers [45–49]. In the case of fibers, three groups have succeeded in obtaining melt-spun fibers with tensile strength of 190–420 MPa from P(3HB) produced by wild-type bacteria. However, the tensile strength of fibers is not sufficient for industrial and medical applications such as fishing line and suture. Recently, we have succeeded in producing strong and flexible fibers with tensile strength of 1.3 GPa and elongation to break of 35% from ultra-high-molecular-weight P(3HB) (UHMW-P(3HB)) produced by recombinant *E. coli* [49]. The strong fibers were processed from amorphous fibers quenched in ice water near the grass transition temperature (Tg) by a method

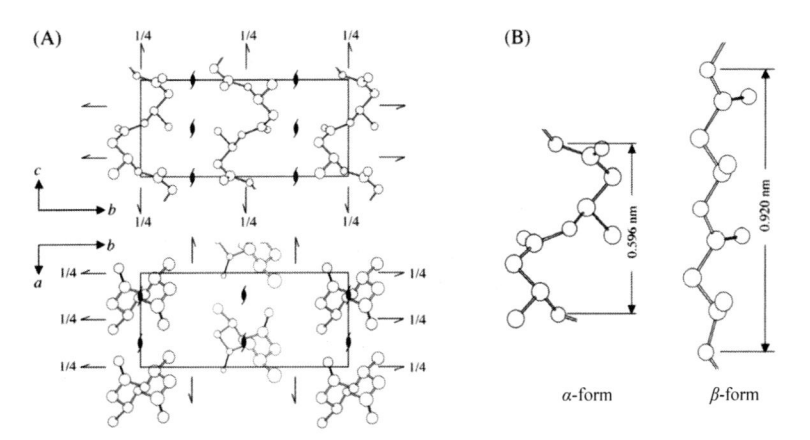

Figure 6 Crystal structure of P(3HB) and two types of molecular conformations: the 2/1 helix conformation (α-form) and the planar zigzag conformation (β-form).

Table 2 Mechanical properties of fibers processed from biodegradable polymers.

Sample	Tensile strength (MPa)	Elongation to break (%)	Young's modulus (GPa)	References
P(3HB)	190	54	5.6	45
	330	37	7.7	46
	310	60	3.8	47
	416	24	5.2	48
	740	26	10.7	50
UHMW-P(3HB)	500	53	5.1	50
	1320	35	18.1	49
P(3HB-co-3HV)	183	7	9.0	57
	210	30	1.8	58
	1065	40	8.0	51
P(4HB)	545	60	0.7	59
P(3HB-co-3HH$_x$)	46	200	–	60
	220	50	1.5	61
	500	50	10.0	62
Poly(lactic acid)	570	35	6.0	63
Poly(glycolic acid)	890	30	8.0	63

combining cold-drawing into ice water and two-step-drawing (a second-drawing) procedure at room temperature. More recently, we developed a new drawing technique (one-step-drawing after growing small crystal nuclei) to obtain the strong fibers from normal molecular weight (commercial grade molecular weight) P(3HB) [50] and poly[(R)-3-hydroxybutyrate-co-(R)-3-hydroxyvalerate] (P(3HB-co-3HV)) [51]. Mechanical properties of PHA fibers together with common plastic fibers are summarized in Table 2.

3. 3. 7 UHMW-P(3HB) fibers

We have developed several new drawing techniques for obtaining strong P(3HB) fibers [49, 52]. The amorphous fibers were obtained by quenching the melt-spun fibers of UHMW-P(3HB) (M_w=5.3×10^6) into ice water. The cold-drawing of amorphous fiber of UHMW-P(3HB) was achieved easily and reproducibly at a temperature below, but near to, the glass transition temperature of 4 °C in ice water with two sets of rolls. The cold-drawn amorphous fibers were kept at room temperature for several minutes to generate the crystal nucleus, and then two-step-drawing was applied by a stretching machine at room temperature. The cold-drawn fibers were easily drawn at very low stress by more than 1000%, but elastic recovery occurred on release from the stretching machine. Accordingly, the annealing procedure is required for fixing the extended polymer chains.

The mechanical properties of cold-drawn, two-step-drawn and annealed fibers of UHMW-P(3HB) are summarized in Table 2. Tensile strength and elongation to break of as-spun fibers was only 38 MPa and 6%, respectively. After cold-drawing for six times in ice water, the tensile strength increased to 121 MPa. Interestingly, the elongation to break was also increased by cold-drawing, indicating that the molecular chains align to the drawing direction and molecular

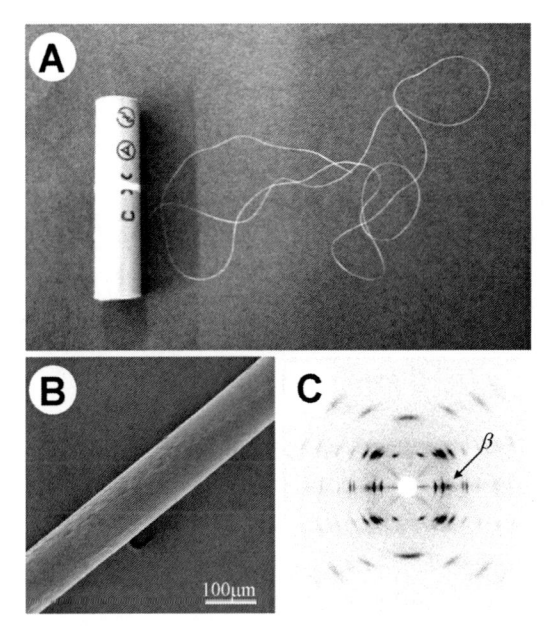

Figure 7 (A) P(3HB) fiber processed by cold-drawn and two-step drawn procedures, (B) scanning electron micrograph, and (C) X-ray fiber diagram.

entanglements decreased by cold-drawing. The tensile strength of two-step-drawn and annealed fibers linearly increased in the ratio of two-step drawing. When the total drawn ratio reached 60 times (cold-drawn for 6 times and two-step-drawn for 10 times), the tensile strength increased to 1,320 MPa. This value is higher than those of poly ethylene, poly propylene, poly(ethylene terephthalate), and poly(vinyl alcohol) of industrial level, and poly(glycolic acid) used as suture. Thus, it was revealed that P(3HB) homopolymer becomes a much attractive material from a view point of mechanical properties (Figure 7). X-ray fiber diagram for P(3HB) fibers is shown in figure 7(C). This diagram includes reflections from both the α-form (2/1 helix conformation) and the β-form (planar zigzag conformation) of P(3HB) simultaneously (figures 6(B) and 7(C)) [53]. β-form is considered as generating from the orientation of free chains in amorphous regions between α-form lamellar crystals [43, 54–56].

3. 3. 8 Structure and function of PHB depolymerase

PHA is a solid polymer with a high molecular weight and incapable of being transported into microorganisms through their cell wall. Therefore, a number of microorganisms such as bacteria and fungi excrete extracellular PHA-degrading enzymes to hydrolyze the solid PHA into water-soluble oligomers and monomer, and after then they utilize the resulting products as nutrients within cells (Figure 1(B)). Thus the biodegradation of PHA materials in natural environments is initiated by the enzymatic hydrolysis reaction with PHA-degrading enzymes.

Many extracellular PHA depolymerases have been purified from different microorganisms and/or characterized [64–69]. The purified PHA depolymerases consisted of a single polypeptide chain and their molecular weights were in the range of 26 000–63 000 (Figure 8) [70–79]. A

Type A

Pseudomonas stutzeri

Type B

Comamonas acidovorans

Comamonas teststeroni

■ Catalytic domain ☐ Linker region ▨ Binding domain ■ Lipase box
(Gly-Xaa-Ser-Xaa-Gly)

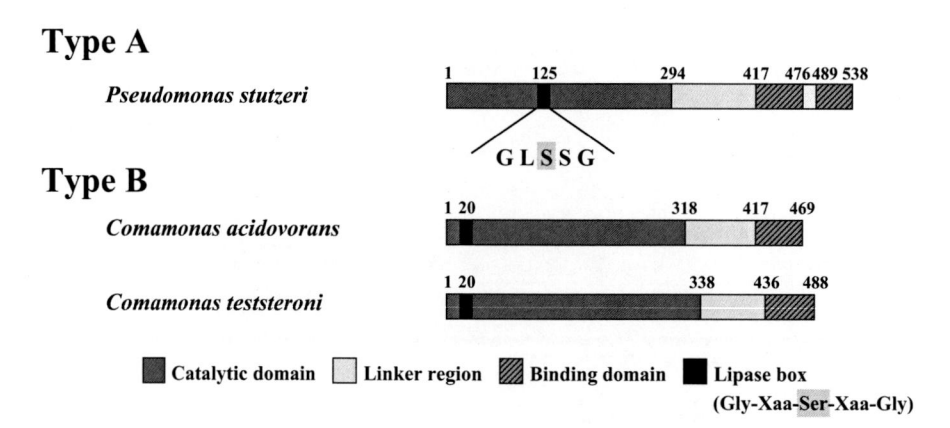

Figure 8 Molecular characterization of PHB depolymerases.

number of extracellular bacterial PHA depolymerase genes have been cloned and analyzed [67, 69, 71, 72, 74–78, 80–88]. Analysis of structural genes of extracellular PHA depolymerases has shown that most enzymes are comprised of N-terminal catalytic domain, C-terminal putative substrate-binding domain, and a linker region connecting the two domains. The catalytic domain contains a lipase box pentapeptide [Gly-Xaa-Ser-Xaa-Gly] as an active site, which is present in almost all known serine hydrolases such as lipases, esterases, and serine-proteases [89–91]. The substrate-binding domain is primarily responsible for the adsorption of the PHA depolymerase to the surface of water-insoluble polyester materials. Based on the hydrolizing and binding assays by using deletion mutants of PHA depolymerases [75, 87, 92–98], each of the catalytic domain and substrate-binding domain functions independently, but these two functional domains and linker region are essential for the enzymatic degradation of PHAs.

The enzymatic hydrolysis experiments using linear and cyclic oligomers of 3HB provide that the PHA depolymerases have the *endo*-hydrolase activity in addition to the *exo*-hydrolase activity [99–101]. In addition, the results demonstrated that the active site of depolymerase recognizes the sequential plural (three or four) monomeric units as substrate for the hydrolysis of single ester bond in polymer chain [95, 97, 99]. Recently, Hisano *et al.* succeeded to obtain the three-dimensional structure of PHA depolymerase from *Penicillium funiculosum* (Figure 9), and demonstrated that P(3HB) chain is bound to the active site through its four sequential monomeric units [102].

3. 3. 9 Industrial production of P(3HB) and its copolymers

P(3HB) and its copolymers, which are accumulated as granules in a wide variety of microorganisms as an intracellular carbon and energy storage material, were initially produced by Imperial Chemical Industries (ICI), UK, by using an industrial-scale fermentation process. Since P(3HB) is very stiff and brittle material because of high crystallinity and secondary crystallization, ICI produced P(3HB-*co*-3HV) from glucose and trademarked as Biopol. In fact, shampoo bottles were produced from P(3HB-*co*-3HV) and were utilized in Europe. However, 3HV units included in the crystal region of P(3HB) homopolymer and the mechanical properties and long-term stability did not

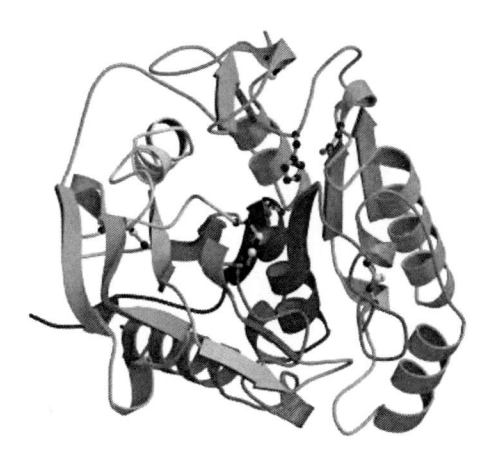

Figure 9 A molecular surface representation of the PHA depolymerase from *Pseudomonas funiculosum.*

improve despite introducing the second monomer unit. In 1995, the process and related patents were bought by Monsanto and subsequently acquired by Metabolix. Mitsubishi Gas Chemicals, Japan, produced about 10 tons of P(3HB) per year by using methanol-oxidizing bacterium. PHB Industrial S/A, Brazil, demonstrates small-scale P(3HB) and P(3HB-*co*-3HV) productions in sugarcane mills and processes some biodegradable agricultural goods. This company has already succeeded to produce bio-ethanol from sugarcane mills and all electronic power used in industrial plants were covered by burning of bagasse from sugarcane. Recently, Chinese companies are very aggressive to produce P(3HB) and its copolymers. ZheJiang TianAn Biologic Materials Co. Ltd., China, produced 2,000 ton per year of P(3HB-*co*-3HV) in high efficiency in collaboration with the Institute of Microbiology affiliated with the Chinese Academy of Sciences.

When 1,4-butandiol is supplied in different amounts as carbon sources, P(3HB-*co*-4HB) is biosynthesized with the ratio of 4HB contents of 5–40 mol%. This copolymer has excellent elongation to break and various thermal properties for various applications. Metabolix, USA, and Tianjin Green Bioscience, China, together with DSM, The Netherlands, announced to produce 50 000 and 10 000 tons per year, respectively [103]. On the other hand, Tepha, USA, produces P(4HB) (TephaFLEX) for medical applications as absorbable sutures and surgical meshes.

The joint venture (Telles) of Archer Daniels Midland Company (ADM) and Metabolix Inc. is now producing polyhydroxyalkanoates (PHA) which are marketed under the commercial name of Mirel^{TM}. This first Mirel production facility, located in Clinton Iowa is now operating and commercial quantities of resin are available for customers globally. Commercial grades of Mirel include injection molding, cast sheet, thermoforming, and film. Mirel can be utilized in a wide variety of applications including food service ware items, packaging, durable goods, compost bags, marine, and in agriculture/horticulture applications where soil biodegradability is an important function of the material. Mirel is biodegradable in natural soil and water environments, home composting systems, and industrial composting facilities, where these facilities are available. The rate and extent of Mirel's biodegradability will depend on the size and shape of the articles made from it. However, like nearly all bioplastics and organic matter, Mirel is not designed to biodegrade in conventional landfills.

P(3HB-*co*-3HHx) is an another important copolymer produced from plant oil. P(3HB-*co*-3HHx) has been firstly trademarked as Nodax by P&G. Recently, Kaneka Corporation is engaged in full-scale development of Keneka PHBH (tentative name) with the target of launching operations in 2010. Production capacity will be 1,000 tons per year and the plan is to eventually raise the annual production capacity to 10 000 tons several years. Increasing the ratio of 3HHx brings out the soft properties and main application is having it processed into film, sheets, foam, injection moldings, fibers, etc (Figure 10).

Four kinds of PHA, P(3HB), P(3HB-*co*-3HV), P(3HB-*co*-4HB) and P(3HB-*co*-3HHx), are produced on a large scale from 1,000 to 50 000 tons per year for commercial exploitation. The annual production of 5,000 tons is estimated to cost from US$3.5 to 4.5 per kg. For reduction of production cost, high cell density cultivation, super-high rate of accumulation of polymers in cell dry weight, continuous fermentation, development of new processing of polymer purification, etc are required. However, thought production cost decreases, the improvement of mechanical properties and the development of new processing methods of materials are also important to open up the market. For being PHA a true new generation polymer, further research and development are needed by the attention of scientist and technician from various disciplines.

Figure 10 Products made from P(3HB-*co*-3HHx) (Photos from Kaneka Corporation, Japan)

References

1. R. Alper et al., *Biopolymers*, **1**, 545 (1963)
2. D. G. Lundgren et al., *J. Bacteriol.*, **89**, 245 (1965)
3. Y. Doi, *Microbial Polyesters*, VCH Publishers, New York (1990)
4. A. J. Anderson, E. A. Dawes, *Microbiol. Rev.*, **54**, 450 (1990)
5. M. Lemoignei, *Bull. Soc. Chim. Biol.*, **8**, 770 (1926)
6. *Biopolymers. vol. 3a (Polyesters I) and 3b (Polyesters II)*, eds A. Steinbüchel, Y. Doi, WILEY-VCH Verlag GmbH, Weinheim (2002)
7. R. W. Lenz, R. H. Marchessault, *Biomacromolecules*, **6**, 1 (2005)
8. J. Lu et al., *Polym. Rev.*, **49**, 226 (2009)
9. K. Taguchi et al., *Biopolymer Handbook, vol. 3a*, Polyesters I, Biological Systems and Biotechnological Production (Part A), edt. WILEY-VCH Verlag GmbH, pp.219 (2002)
10. S. C. Slater et al., *J. Bacteriol.*, **170**, 4431 (1988)
11. P. Schubert et al., *J. Bacteriol.*, **170**, 5837 (1988)
12. T. Fukui et al., *J. Bacteriol.*, **180**, 667 (1998)
13. B. H. Rehm et al., *J. Biol. Chem.*, **273**, 24044 (1998)
14. K. Taguchi et al., *FEMS Microbiol. Lett.*, **176**, 183 (1999)
15. Q. Ren et al., *J. Bacteriol.*, **182**, 2978 (2000)
16. C. T. Nomura et al., *Appl. Environ. Microbiol.*, **70**, 999 (2004)
17. G. J. M. De Koning, P. J. Lemstra, *Polymer*, **34**, 4089 (1993)
18. M. Scandola et al., *Macromol. Chem. Rapid Commun.*, **10**, 47 (1989)
19. P. A. Holmes, *Developments in crystalline polymers. vol. 2.*, ed D. C. Bassett, 1, Elsevier Applied Science, London and New York (1988)
20. S. Nakamura et al., *Macromolecules*, **25**, 4237 (1992)
21. M. Pizzoli et al., *Macromolecules*, **24**, 4755 (1994)
22. T. Tsuge, *J. Biosci. Bioeng.*, **94**, 579 (2002)
23. Y. Saito, Y. Doi, *Int. J. Biol. Macromol.*, **16**, 99 (1994)
24. H. Matsusaki et al., *Biomacromolecules*, **1**, 17 (2000)
25. T. Fukui et al., *J. Bacteriol.*, **179**, 4821 (1997)
26. T. Fukui et al., *Appl. Microbiol. Biotechnol.*, **49**, 333 (1988)
27. T. Tsuge et al., *Sci. Technol. Adv. Mater.*, **5**, 449 (2004)
28. P. Kahar et al., *Polym. Degrad. Stab.*, **83**, 79 (2004)
29. M. Akiyama et al., *Polym. Degrad. Stab.*, **80**, 183 (2003)
30. S. Kusaka et al., *Appl. Microbiol. Biotechnol.*, **47**, 140 (1997)
31. S. Kusaka et al., *J. Macromol. Sci. Pure Appl. Chem.*, **A35**, 319 (1998)
32. F. Wang et al., *Appl. Environ. Microbiol.*, **63**, 4765 (1997)
33. P. Kahar et al., *Polym. Degrad. Stab.*, **87**, 161 (2005)
34. J. Agus et al., *Polym. Degrad. Stab.*, **91**, 1645 (2006)
35. J. Choi et al., *Biotechnol. Bioproc. Eng.*, **9**, 196 (2004)
36. J. Agus et al., *Polym. Degrad. Stab.*, **91**, 1138 (2006)
37. M. Yokouchi et al., *Polymer*, **14**, 267 (1973)
38. K. Okamura, R. H. Marchessault, *Conformation of biopolymers, vol. 2*, ed G. N. Ramachandra, Academic Press, New York, 709 (1967)
39. G. J. M. De Koning, P. J. Lemstra, *Polymer*, **35**, 4598 (1994)
40. P. J. Barham, A. Keller, *J. Polym. Sci. Polym. Phys. Ed.*, **24**, 69 (1986)
41. T. Iwata, Y. Doi, *Macromolecular Symposia*, **224**, 11 (2005)

42. S. Kusaka et al., *Int. J. Biol. Macromol.*, **25**, 87 (1999)
43. Y. Aoyagi et al., *Polym. Degrad. Stab.*, **79**, 209 (2003)
44. T. Iwata et al., *Polym. Degrad. Stab.*, **79**, 217 (2003)
45. S. A. Gordeyev, Y. P. Nekrasov, *J. Mater. Sci. Lett.*, **18**, 1691 (1999)
46. G. Schmack et al., *J. Polym. Sci.: Part B: Polym. Phys.*, **38**, 2841 (2000)
47. H. Yamane et al., *Polymer*, **42**, 3241 (2001)
48. Y. Furuhashi et al., *Polymer*, **45**, 5703 (2004)
49. T. Iwata et al., *Macromol. Rapid Commun.*, **25**, 1100 (2004)
50. T. Tanaka et al., *Polym. Degrad. Stab.*, **92**, 1016 (2007)
51. T. Tanaka et al., *Macromolecules*, **39**, 2940 (2006)
52. T. Iwata et al., *Macromolecules*, **17**, 5789 (2006)
53. T. Iwata et al., *Biomacromolecules*, **6**, 1803 (2005)
54. T. Iwata, T. Tanaka, *Nippon Gomu Kyokaishi*, **81**, 358 (2008)
55. W. J. Orts et al., *Macromolecules*, **23**, 5368 (1990)
56. T. Iwata, *Macromol. Biosci.*, **5**, 689 (2005)
57. T. Ohura et al., *Polym. Degrad. Stab.*, **63**, 23 (1999)
58. T. Yamamoto et al., *Int. Polym. Processing*, **XII**, 29 (1997)
59. D. P. Martin, S. F. Williams, *Biochem. Eng. J.*, **16**, 97 (2003)
60. E. B. Bond, *Macromol. Symp.*, **197**, 19 (2003)
61. Y. Jikihara et al., *SEN'I GAKKAISHI*, **62**, 115 (2006)
62. T. Tanaka et al., *Fiber Preprints*, Japan, **62**, 103 (2007)
63. *Polymer Handbook*, eds J. Brandrup, E. H. Immergut, E. A. Grulke, John Wiley & Sons, Inc., New York (1999)
64. D. Jendrossek et al., *J. Environ. Polym. Degrad.*, **1**, 53 (1993)
65. K. Mukai et al., *Polym. Degrad. Stab.*, **41**, 85 (1993)
66. T. Tanio et al., *Eur. J. Biochem.*, **124**, 71 (1982)
67. K. Yamada et al., *Int. J. Biol. Macromol.*, **15**, 215 (1993)
68. K. Nakayama et al., *Biochim. Biophys. Acta*, **827**, 63 (1985)
69. M. Uefuji et al., *Polym. Degrad. Stab.*, **58**, 275 (1997)
70. D. Jendrossek et al., *Eur. J. Biochem.*, **218**, 701 (1993)
71. K. Kita et al., *Appl. Environ. Microbiol.*, **61**, 1727 (1995)
72. B. H. Briese et al., *J. Environ. Polym. Degrad.*, **2**, 75 (1994)
73. D. Jendrossek et al., *Can. J. Microbiol.*, **41**, 160 (1995)
74. D. Jendrossek et al., *J. Bacteriol.*, **177**, 596 (1995)
75. K. Kasuya et al., *Appl. Environ. Microbiol.*, **63**, 4844 (1997)
76. B. Klingbeil et al., *FEMS Microbiol. Lett.*, **142**, 215 (1996)
77. T. Saito et al., *J. Bacteriol.*, **171**, 184 (1989)
78. M. Shinomiya et al., *FEMS Microbiol. Lett.*, **154**, 89 (1997)
79. K. Mukai et al., *Polym. Degrad. Stab.*, **43**, 319 (1994)
80. T. Watanabe et al., *J. Bacteriol.*, **174**, 408 (1992)
81. T. Kobayashi et al., *J. Environ. Polym. Degrad.*, **7**, 9 (1999)
82. K. Kita et al., *Biochim. Biophys. Acta*, **1352**, 113 (1997)
83. M. Takeda et al., *J. Biosci. Bioeng.*, **90**, 416 (2000)
84. A. Schirmer, D. Jendrossek, *J. Bacteriol.*, **176**, 7065 (1994)
85. A. Schirmer et al., *Can. J. Microbiol.*, **41** (Suppl. 1), 170 (1995)
86. U. Schöber et al., *Appl. Environ. Microbiol.*, **66**, 1385 (2000)
87. M. Nojiri, T. Saito, *J. Bacteriol.*, **179**, 6965 (1997)

88. K. Kasuya et al., *Int. J. Biol. Macromol.*, **33**, 221 (2003)
89. K. E. Jaeger et al., *Annu. Rev. Microbiol.*, **53**, 315 (1999)
90. K. E. Jaeger et al., *FEMS Microbiol. Rev.*, **15**, 29 (1994)
91. K. E. Jaeger et al., *Appl. Environ. Microbiol.*, **61**, 3113 (1995)
92. A. Behrends et al., *FEMS Microbiol. Lett.*, **143**, 191 (1996)
93. B. H. Briese, D. Jendrossek, *Macromol. Symp.*, **130**, 205 (1998)
94. T. Fukui et al., *Biochim. Biophys. Acta*, **952**, 164 (1988)
95. T. Hiraishi et al., *Biomacromolecules*, **1**, 320 (2000)
96. K. Kasuya et al., *Int. J. Biol. Macromol.*, **24**, 329 (1999)
97. T. Ohura et al., *Appl. Environ. Microbiol.*, **65**, 189 (1999)
98. M. Shinomiya et al., *Int. J. Biol. Macromol.*, **22**, 129 (1998)
99. B. M. Bachmann, D. Seebach, *Macromolecules*, **32**, 1777 (1999)
100. H. Brandl et al., *Can. J. Microbiol.*, **41** (Suppl. 1), 180 (1995)
101. Y. Shirakura et al., *Biochim. Biophys. Acta*, **880**, 46 (1986)
102. T. Hisano et al., *J. Mol. Biol.*, **356**, 993 (2006)
103. G. Q. Chen, *Plastics from Bacteria: Natural Functions and Applications*, Microbiology Monographs 14, ed. G. Q. Chen, Springer, pp.121 (2010)

3. 4 Poly (trimethylene terephthalate, PTT)
Mureo Kaku

3. 4. 1 Introduction
Since entering the 21st century, it has become quite obvious the environment surrounding us is getting worse and there are serious signs that our daily lives are in great jeopardy. Revolutionary change of life style in developed countries is considered to be a contributing factor to this serious situation. In the 20st century, the way we lived changed dramatically compared to any other time in human history. "Sustainability" for our earth is becoming increasingly doubtful but establishing a sustainable society is critical for the future of the human race.

There are three serious problems, the so-called "Mega Trend", facting us now, 1) Shortage of food due to the rapid increase of population especially in developing countries, 2) Rapid increased consumption of petroleum which is a depletable natural resource, resulting in higher oil prices in the long term, and 3) Global warming and abnormal weather patterns. Reduction of greenhouse gas (GHG) such as CO_2 is required. Each of these problems is complicated and they are somehow related. DuPont belives these problems are solvable and advanced bio-technology is the key to solving them.

To respond to these mega trends and offer speedy and concrete solutions to society, DuPont has decided to develop biotechnology in the area of 1) food, 2) bio-fuel and 3) bio-based material to serve society's needs for a sustainable future.

Genetic modified technology developed to increase crops with additional insect resistance, drought resistance and also offering higher nutrition is considered to be as one of the most promising methods to increase the production of crops to solve the food shortage issue. The technology derived from agriculture can be applied to the production of bio-fuel and bio-material which is believed to lower the environmental footprint and thus ease the exhaustion of natural resources issue. This technology is called as "industrial biotechnology" where biomass is used as the starting raw material but with advanced bio-technology, the process differs from traditional fermentation which relies on readily available enzymes and microbes. Under the newly developed bio-catalyst, the metabolic pathway of microbes is modified and rebuilt such to the way that all biomass is consumed and converted to the targeted product. Of course, the efficiency of bio-catalyst will be further improved and target yield will increase.

DuPont has established its new "2015 Sustainability Goals", which tie DuPont's business growth to the development of products that have positive environmental impacts. In these new goals, DuPont is committed to:
(1) Doubling its investment in research and development programs with direct and quantifiable environmental benefits for customers.
(2) Growing annual revenues at least $2 billion from products that significantly contribute to

the reduction of GHG emissions for customers.
(3) Doubling revenues from non-depletable resources to at least $8 billion.

In the development of bio-based material, DuPont has established the following ground rules:
(1) The targeted material must have unique functionality.
(2) The targeted material is from the bio-route and lower environmental footprint can be achieved.
(3) Costs from the bio-route will be competitive to compare with those from chemical processes.

3. 4. 2   Bio-1,3 propanediol (Bio-PDO™)

Consistent with these goals, 1,3-propandiol (PDO) is the first example selected by DuPont to be produced from the bio-process and commercialized.

In the nature, PDO is produced in two steps, glucose is converted to glycerol by yeast, then glycerol is further processed to PDO by bacterium, but the yield is very low and is not suitable for industrial use (Figure 1).

It is critical to develop a process in which glucose can be converted to PDO without generating by-product. By utilizing advanced bio-technology, DuPont has developed a bug (Figure 2), whose metabolic system is controlled to convert glucose only to PDO [1, 2].

In November 2006, DuPont has formed a joint venture with Tate & Lyle, known as one of

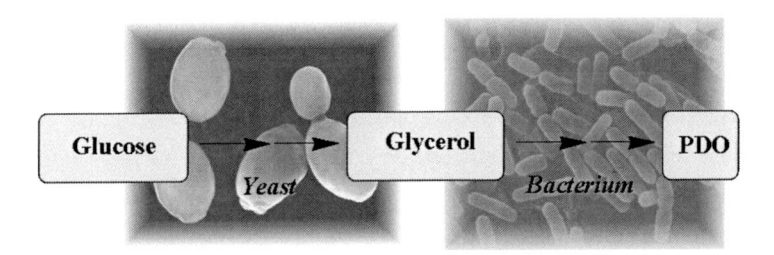

Figure 1   PDO production process in nature

- Glucose transport
- Glycerol metabolism
- Glycolysis
- Entner-Doudoroff
- Pentose phosphate
- TCA cycle
- Respiration
- Amino acid biosynthesis
- Anapleurotic reactions
- Global regulators

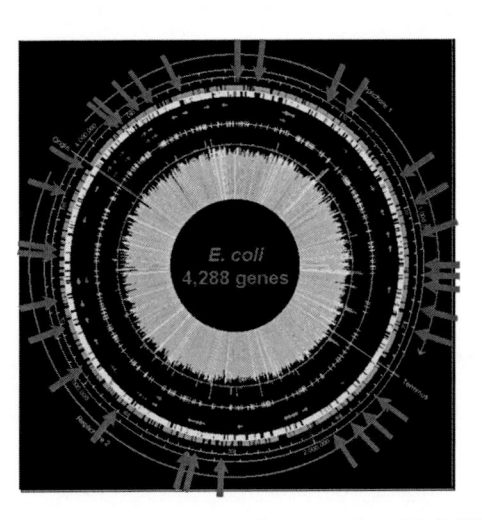

Figure 2   Modification of E.coli to obtain necessary function to produce PDO [3]

88   Bio–Based Polymers

Figure 3   DuPont Tate & Lyle BioProducts Loudon Plant

the largest food and corn sugar companies to commercialize Bio-PDO™. DuPont Tate & Lyle BioProducts with a capacity of 45 000 tons is now in full operation at Loudon, Tennessee (Figure 3).

With the newly developed biocatalyst concept, this plant is one of the largest and most efficient plants to produce chemicals from fermentation process. The success of this plant suggests that bio-technology can contribute to both business and the environment in a sustainable manner.

3. 4. 3   Poly (trimethylene terephthalate, PTT)

Poly (trimethylene terephtalate) (PTT) can be produced by standard condensation polymerization of PDO and terephthalic acid [4] (Scheme 1). Similar to polyethylene terephthalate (PET) and Poly (butylene terephthalate) (PBT), PTT is a crystalline thermoplastic polyester. DuPont that commercialized PTT under the trade name of Sorona® is using Bio-PDO® as a raw material, and its bio-content is analyzed as 35.9%, meeting the criteria for of bio-polymers set by the Japan BioPolymer Association (JBPA). It is listed on the positive list of No. B08003 (Figure 4).

Figure 5 shows the thermal properties of PTT, PET and PBT are shown. It is obvious that both the melting point and glass transition temperature of PTT are between PET and PBT following the number of carbons in glycol moiety. Although the thermal properties of PTT fall between PET and PBT (Figure 5), the molecular structure of PTT is quite unique and different from both PET and PBT which is in linear mode (Figure 6). The kink existing in PTT is derived from the unique angle of the PDO molecule [5], which provides the excellent and unusual physical properties of PTT that cannot be expected from PET and PBT.

3. 4. 4   Sorona® polymer for fiber applications

Figure 7 shows a comparison of the mechanical properties of fiber made with Sorona® polymer with PET, Nylon 6,6 and PBT. Different from other fibers, the fiber made with Sorona® polymer has a soft touch with comfortable stretch and recovery properties that are suitable for carpet and textile applications as summarized in Table 1.

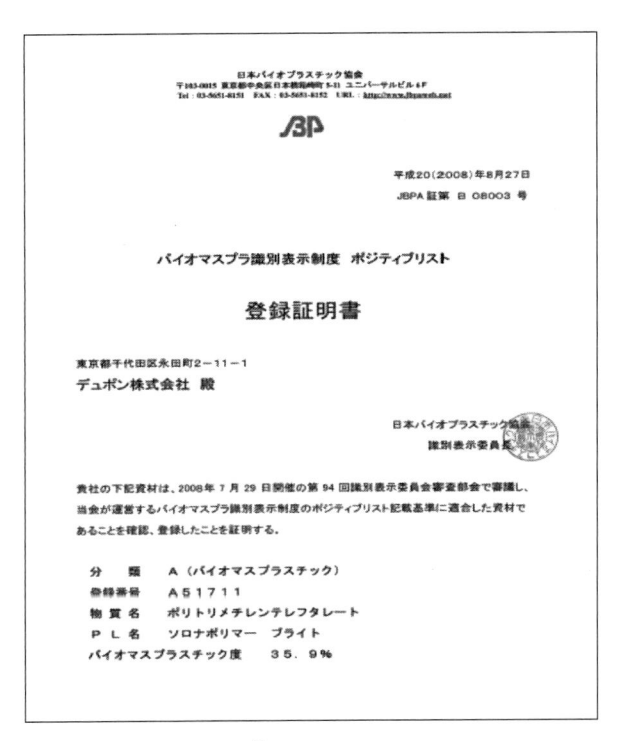

Figure 4 Sorona® polymer registration in JBPA

$$RO_2C-\bigcirc-CO_2R + HO(CH_2)_xOH \xrightarrow{-ROH} [OC-\bigcirc-CO_2(CH_2)_xO]_n$$

R=H: TPA, R=CH$_3$:DMT   X=2: PET, X=3: PTT, X=4: PBT

Scheme 1 Condensation polymerization of PTT

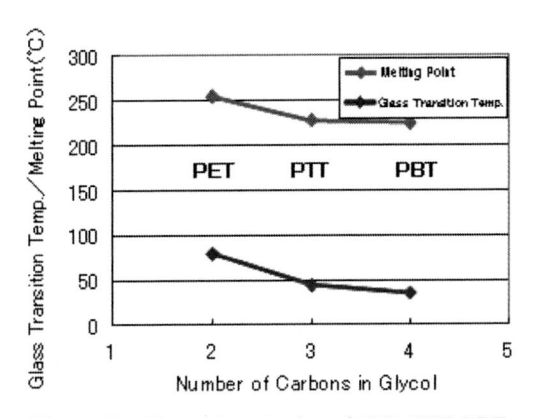

Figure 5 Thermal properties of PET, PTT, PBT

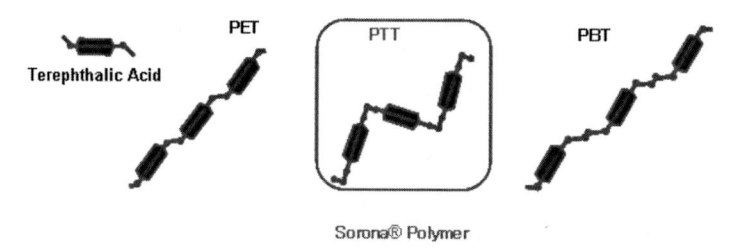

Figure 6 Unique molecular structure of PTT

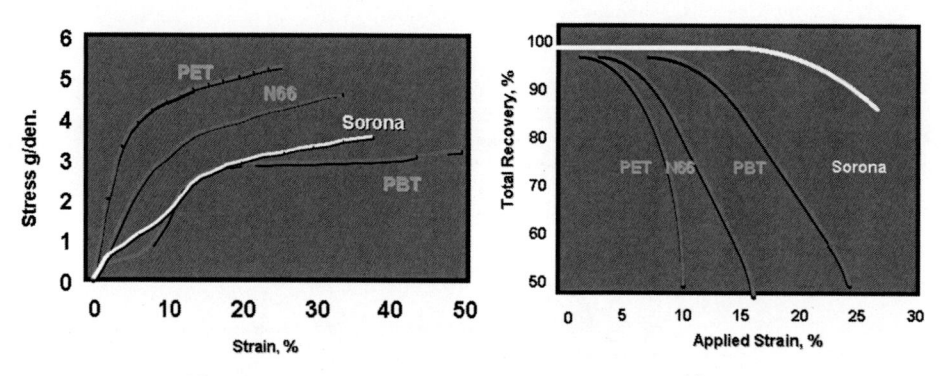

Figure 7 Unique physical characteristics of Sorona® fiber

Table 1 Characteristic of Sorona® fiber

Stain resistance
Softness
Vibrant color
Stretch & recovery
UV & chlorine resistance
Easy care
Dimensional stability

Fiber made with Sorona® polymer exhibits very unique characteristics in comparison to Nylon fiber which is a major material used in carpets. In order to improve dirt resistance, it is necessary to apply fluorinated material to Nylon fibers. But as shown in Figure 8, fiver made with Sorona® polymer has inherent dirt repellency based on its unique molecular structure and does not require any treatment such as F-chemical coating which is not only costly but also not environmentally friendly.

Since the glycol portion of Sorona® polymer is from bio-PDO which is renewably resourced, it was revealed that Sorona® polymer emitted 55% less GHG than that of Nylon polymer (Figure 9) in its LCA assessment.

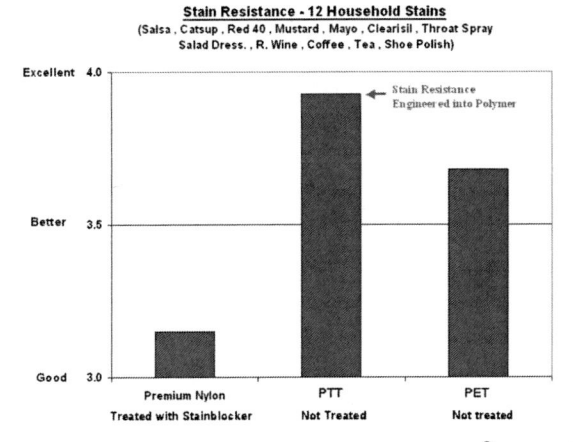

Figure 8 Comparison of stain resistance between Sorona® and Nylon fibers

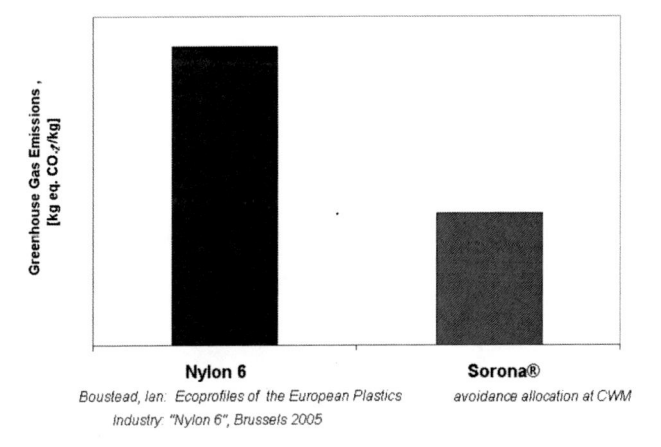

Figure 9 Comparison of Sorona®/Nylon for GHG emission

3. 4. 5 Sorona® polymer for injection mold applications

As a functional engineering polymer, injection mold application is also a promising area for Sorona® polymer. Under the brand name of Sorona® EP, various additives can be compounded with Sorona® polymer to further strengthen its properties. In Table 2 the physical properties of two glass fiber reinforced grades, Sorona® 3015G (containing 15% glass fiber) and 3030G (containing 30% glass fiber) of Sorona® EP are shown.

Sorona® is a bio-based polymer but is not biodegradable. Its physical properties, mechanical properties, hydro-stability and durability are more superior to compare with other bio-based polymers such as starch and polylactic acid (PLA) (Figure 10) acid. Sorona® EP inherent the properties derived from Sorona® polymer, it has properties comparable to other engineering plastics and is suitable for automotive and electronic parts applications [6].

Similar to other engineering polymers, all mold machines with in-line screw can be used to process Sorona® EP. The melting point of Sorona® EP is 227°C, close to PBT resin (225°C), and the recommended melted resin temperature is 240–260°C at a die temperature of 80–110°C.

Table 2 General properties of Sorona® EP glass fiber reinforced grades

Properties Test	Testing Method	unit		Sorona® 3015G 15% Glass fiver	Sorona® 3030G 30% Glass fiver
Mechanical Properties					
Tensile Strength	ISO 527-1, 2	MPa	23°C	123	162
Elongation	ISO 527-1, 2	%	23°C	3.0	2.5
TensileModulus	ISO 527-1, 2	MPa	23°C	6,200	10,400
Flexural Strength	ISO 178	MPa	23°C	190	245
Flexural Modulus	ISO 178	MPa	23°C	5,700	9,600
Izod (with notch)	ISO 179	kj/m2	−30°C	6.0	9.0
		kj/m2	23°C	5.5	9.0
Izod (no notch)	ISO 179	kj/m2	−30°C	30	45
		kj/m2	23°C	30	50
Thermal Property					
Melting Point	ISO 11357	°C	°C	227	227
Others					
Gravity	ISO 1183			1.40	1.56
Mold Shrinkage	Die Temp. 80°C	%	Flow	0.5	0.3
(2 mm thick)		%	Perpendicular	0.7	0.8
Mold Shrinkage	Die Temp. 110°C	%	Flow	0.6	0.4
(2 mm thick)		%	Perpendicular	0.8	0.8

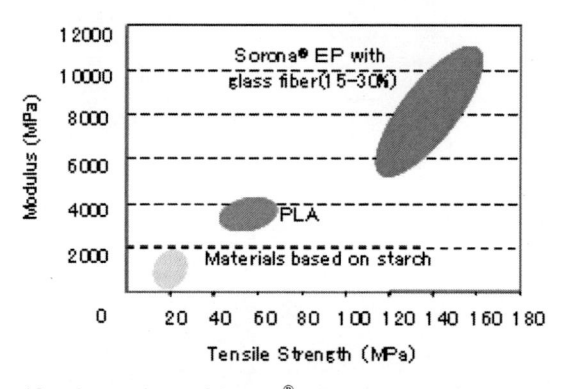

Figure 10 Comparison of Sorona® EP and PLA and starch materials

Unlike with PET and PBT materials, Sorona® EP shows dimensional stability due to less absorption of moisture, and dimension change only occurs during the injection mold process (Figure 11). Compared with PBT parts, the dimension change of Sorona® EP is much smaller in both the injection molding process and post treatment. It is also confirmed that the warpage of the Sorona® part is smaller than that of PBT (Figure 12).

During the injection process, Sorona® EP material solidifies more slowly than PBT material, it allows the Sorona® EP mold part to have sufficient die transcription time at the pressure holding stage thus offering superior surface appearance (Figure 13). This unique characteristic is considered that it is useful when applying Sorona® EP molded parts to reflectors (Figure 14) and door handle applications (Figure 15).

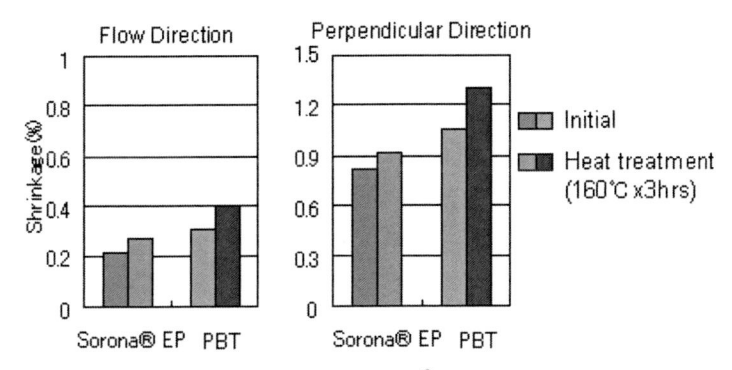

Figure 11 Shrinkage of Sorona® EP 3030G and PBT
(100 mm × 2 mm thick plate)

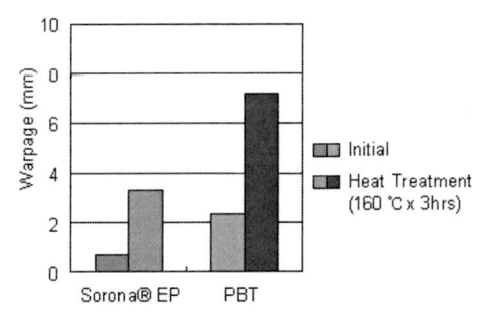

Figure 12 Warpage of Sorona® EP 3030G and PBT
(100 mm × 100 mm × 2 mm plate)

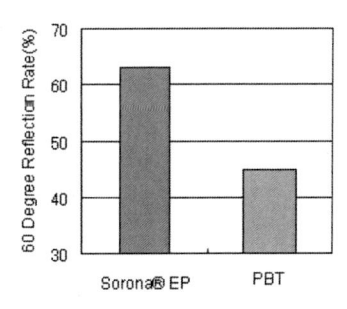

Figure 13 Comparison of surface reflection rate of Sorona® EP 3030G and PBT

Figure 14 Sorona® EP potential applications in reflectors

Figure 15 Sorona® EP potential application in door handles

3. 4. 6 Summary

With the aim of becoming world's most dynamic scientific company creating sustainable solutions to provide better, safer, and healthier living for people everywhere, DuPont has started to apply newly developed bio-technologies to its the business in the area of food, bio-fuel and bio-based materials.

At DuPont & Tate & Lyle BioProducts, the production of Bio PDO™ has started in November 2006. Since then products containing Bio-PDO™ have been commercialized in many areas including cosmetics [7], polyurethanes and unsaturated polyesters. Due to its unique properties and reduced environmental footprint, PTT derived from Bio-PDO™ is gaining popularity and business in carpet and apparel applications is growing well. Following the expanded recognition of Sorona® polymer, Sorona® EP is starting to be adapted for automotive applications.

The development of bio-material has just begun and it is believed that it will offer one of the most powerful solutions to solve the problem of exhausted fossil resources and global warming issues. Recently these bio-materials have been receiving favorable attention in automotive and electronic ECO products because they suitably complement the concept of bio-based material. But in evaluating the contribution of bio-based material to the environment, it is necessary to understand the origin of raw materials, non-food supply chain, production volume and LCA accurately. It is also crucial that the bio-material has unique characteristics and the capability to become a commodity, so its impact on the environment can be maximized. In order to expand the usage of bio-based materials to be the standard of our daily life, it is highly recommended that more bio-based materials are developed and commercialized with unique properties and competitive cost.

Reference

1. EP 1204775, EP1076708, WO0112833, WO111070
2. S. E. Manahan, *Environmental Chemistry*, P503, CRC Press (2005)
3. US 6331264, US 6325945, US 6281325, WO 0158980
4. J. V. Kurian, "Natural fibers, Biopolymers and Biocomposites", Chapter 15, P487-525, CRC Press (2005)
5. A. K. Mohanty, W. Liu, L.T. Drzal. M. Misra, J.V. Kurian, R.W. Miller and N. Strickland, American Institute of Chemical Engineers (AIChE) 2003 Annual Conference, November 16-21, 2003, San Francisoco, California, Paper Number [491e]
6. H. Sumi, *Journal of Society of Automotive Engineers of Japan*, **63**(4) (2009)
7. Household and Personal Care Today nr/3/2007, P30-31, Am HPC's customer publication

Chapter 4
New Polymerization Methods for Bio-based Polymers

4.1 New polymerization methods for bio-based polymers from renewable vinyl monomers

Kotaro Satoh and Masami Kamigaito

4.1.1 Introduction

From the viewpoint of environmentally benign and sustainable chemistry, the use of renewable resources have been attracting much attention as alternatives for traditional petrochemical raw materials from fossil resources [1]. Especially, the raw materials for synthetic polymers abundantly obtained or derived from natural plants are now being studied in place of the general-purpose petrochemical monomers [2–8]. Although this viewpoint is still controversial in some aspects, the suitable and judicious applications of specific or complicated structures originating from natural products is definitely beneficial for developing high performance or functional bio-based polymeric materials. Whereas most bio-based polymers have been prepared from starch-originated or plant-derived monomers via step-growth polymerization, numerous kinds of vinyl monomers for chain-growth polymerizations can be derived from natural products and sometimes occur in nature. Unfortunately, most of the polymers produced through vinyl polymerization would be non-biodegradable. However, it may somehow meet the current world demand that we have to reduce the emission of carbon dioxide by reducing the use of fossil oil. Actually, vinyl compounds abundantly obtained or derived from natural plants are now being studied to replace petroleum-derived common monomers for developing novel bio-based polymeric materials [4].

Meanwhile, a large numbers of polymerization techniques for vinyl monomers have been developed since starting the era of petrochemical industry [9]. Recent years have further witnessed a large number of polymerization systems or catalysts for petrochemical-derived vinyl monomers that can precisely control the primary structures of the producing polymers. The controls of primary structures during chain-growth vinyl polymerization, such as molecular weight, stereochemistry, and monomer sequence, should lead to development of novel functional polymers that would rival natural macromolecules with a uniform molecular weight, stereoregularity, and regulated monomer sequence. Among them, living or controlled polymerization of vinyl monomers, which can precisely control the molecular weights and the terminal groups, has opened a new field of precision polymer synthesis and has been further applied to a wide variety of functional materials based on the controlled polymer structures.

This chapter describes recent developments in controlled/living polymerizations of naturally occurring or derived vinyl monomers, which may lead to creating novel bio-based polymers with well-defined structures. Especially, we will here describe several natural and/or renewable vinyl monomers compared to the petrochemical common vinyl monomers such as olefins, styrenes, and (meth)acrylates as shown in Figure 1.

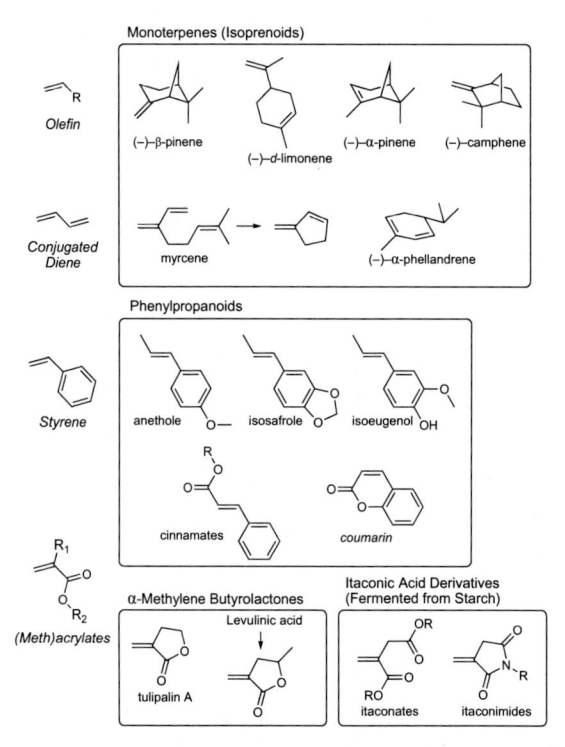

Figure 1 Renewable Vinyl Monomers Occurring or Derived from Natural Plants

4. 1. 2 Controlled/Living polymerization of petrochemical vinyl monomers

From the breakthrough discovery of living anionic polymerization by Szwarc in the 1950s [10], living ionic polymerizations have made noticeable progress in controlling polymer structures and compositions and have been used for the synthesis of various polymers with well-defined architectures. In the 1980s, the living cationic polymerizations of vinyl ethers and styrene derivatives were developed by the combination of a protonic acid (HA) and a Lewis acid (MtX_n), in which the former (or its adduct with a monomer) serves as an initiator to form a dormant carbon-halogen or related covalent linkage (∼∼C-A), and the latter as a catalyst or an activator that assists the formation of a carbocation from the initiator and the dormant polymer end and thereby triggers the controlled propagation (Scheme 1) [11,12]. A key to accomplishing the living cationic polymerization is the extension of the lifetime of the growing cations through the reversible conversion of an active carbocation to a dormant state. It is required to select a moderately nucleophilic and dissociative counteranion (A⁻) and a Lewis acid (MtX$_n$) with a moderate acidity. These produced polymers exhibit the features of living polymerization; *i.e.*, controlled molecular weights determined by the initial ratio of the initiator to monomer, narrow molecular weight distributions, and well-defined end groups among others.

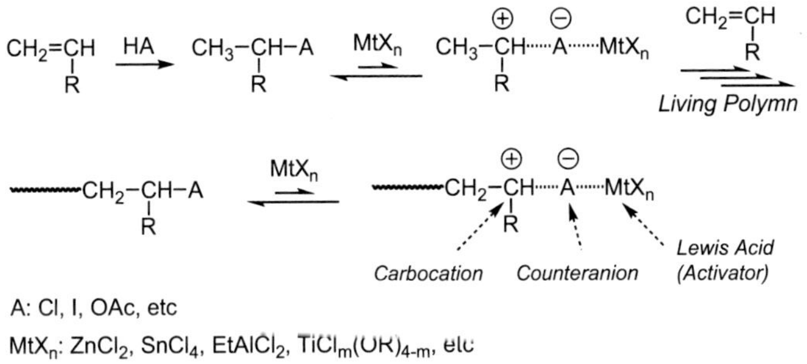

Scheme 1 Living Cationic Polymerization of Vinyl Monomers

In the mid 1990s, a great deal of progress was made in controlled/living radical polymerization which permits control of the primary structures of the polymers for numerous types of radically-polymerizable vinyl monomers even with polar and functional groups such as carbonyl, carboxylic acid, hydroxyl, and amine. Until then, the precise control of chain-growth radical polymerization had been considered beyond one's reach because of the recombination and disproportionation reactions of neutral and highly active radical-growing species. To overcome this problem, the dormant-active concept was conducted at the growing chain end for taming radical polymerization as in the living cationic polymerization, of which covalent species as the dormant species was also reversibly activated or transformed into the growing radical by certain stimuli. The controlled/living radical polymerization can now be mainly categorized into three processes, the nitroxide-mediated polymerization (NMP) [13–15], the metal-catalyzed living radical polymerization or atom transfer radical polymerization (ATRP) [16–22], and the reversible addition-fragmentation chain transfer (RAFT) or macromolecular design via interchange of xanthates (MADIX) polymerization [23–26]. Scheme 2 shows the three representative polymerizations, all of which are based on the transient, fast, and reversible activation of the dormant species into the growing radical species. The key to the molecular weight control mainly lies in the predominant formation of the dormant species from the growing radical species as well as the establishment of an extremely fast activation-deactivation process relative to the propagation process.

In the aforementioned polymerization systems, the judicious initiating systems are generally chosen or designed depending on the monomers for attaining the fine control of the precise polymerization systems, because the propagating rate, nature of the growing species, and sometimes the strength of the covalent dormant species should change with the monomer types. By evolving the polymerization systems, one may polymerize less-noticed naturally-occurring or

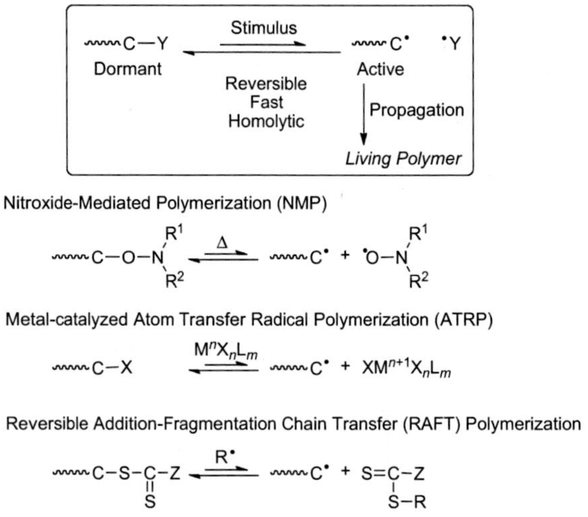

```
         ┌─────────────────────────────────────────────┐
         │                    Stimulus                 │
         │   ∼∼∼C—Y    ⇌           ∼∼∼C•   •Y          │
         │   Dormant               Active              │
         │             Reversible                      │
         │                Fast    │ Propagation        │
         │             Homolytic  │                    │
         │                        ↓                    │
         │                        Living Polymer       │
         └─────────────────────────────────────────────┘
```

Nitroxide-Mediated Polymerization (NMP)

$$\sim\!\!\sim\!\!\sim\text{C–O–N}\overset{R^1}{\underset{R^2}{}} \xrightleftharpoons{\Delta} \sim\!\!\sim\!\!\sim\text{C}^\bullet + {}^\bullet\text{O–N}\overset{R^1}{\underset{R^2}{}}$$

Metal-catalyzed Atom Transfer Radical Polymerization (ATRP)

$$\sim\!\!\sim\!\!\sim\text{C–X} \xrightleftharpoons{M^n X_n L_m} \sim\!\!\sim\!\!\sim\text{C}^\bullet + XM^{n+1}X_n L_m$$

Reversible Addition-Fragmentation Chain Transfer (RAFT) Polymerization

$$\sim\!\!\sim\!\!\sim\text{C=S–C–Z} \xrightleftharpoons{R^\bullet} \sim\!\!\sim\!\!\sim\text{C}^\bullet + \text{S=C–Z}$$
$$\qquad\quad\;\;\overset{\|}{S} \qquad\qquad\qquad\qquad\quad\;\; \overset{|}{S\text{–R}}$$

Scheme 2 Representative Living Radical Polymerizations of Vinyl Monomers

-derived monomers into well-defined structures in order to develop novel bio-based polymers or copolymers.

4. 1. 3 Polymerizations of naturally-occurring olefins (Terpenes)
Cationic polymerization of terpenes for bio-based cycloolefin polymer

Alicyclic hydrocarbon polymers (cycloolefin polymers) are now extensively utilized especially in the optoelectronic fields, not only because of their rigid backbone having a high service temperature and high mechanical strength, but also of their low dielectric constant, nonhygroscopicity, and good transparency [27–30]. The vinyl addition polymerization of cyclic olefins is one of the most promising methods that can readily produce amorphous polymers retaining the cyclic structures, while another approach has recently been developed and applied on an industrial scale by ring-opening metathesis polymerization of polycyclic olefins or coordination polymerization of cyclic olefins in the petroleum fraction [31–35].

Meanwhile, a wide range of alicyclic polymerizable olefins can be found in naturally-occurring terpenes, such as limonene, terpinen, phellandrene, and pinenes. Some terpenes found in the natural products of plants are non-polar and mono- or bicyclic olefins with their characteristic structures sometimes containing chirality [36,37]. Since the 1950s, the homopolymerizations of terpenes have been intensively investigated via carbocationic mechanism [38–40]. However, in general, those have resulted in low molecular weight polymers with inefficient strength and a low service temperature ($T_g < 100\ °C$), which limited the commercial utilization of poly(terpene)s or of their hydrogenated forms not for polymer materials, but for additives, such as tackifiers in adhesives, or modifiers in molding compounds [27]. The cationic homopolymerizations of naturally-occurring terpenes and the subsequent hydrogenations will lead to developing bio-based cycloolefin polymers having good properties for practical use, especially their thermal properties with a high service temperature and durability.

Pinenes are the main component of the abundant and cheap pine tree oil, which is called turpentine and still industrially produced worldwide over 330 thousand metric tons per year [41,42]. β-Pinene, the second most common component of turpentine, must be one of the most promising sustainable resources for the purpose of cycloolefin polymers in terms of not only availability but also the polymerizability by cationic polymerization. The cationic polymerization of β-pinene has also been intensively investigated via addition and ring-opening cationic mechanism [38]. The β-pinene polymerization had also resulted in low molecular weight polymers with a low glass transition temperature, among which the highest T_g value ever reported was only 65°C by Kennedy et al. [43]. The controlled/living cationic polymerization of β-pinene was reported by Higashimura and Deng et. al. in 1997 with $TiCl_2(OR)_2$ as the catalyst in conjunction with the HCl adduct of an alkyl vinyl ether as the initiator (Scheme 3) [44]. The living polymerization of β-pinene was also successfully used for preparing various block and graft copolymers [45–50]. However, even with the Ti-based system, the control of the molecular weight of poly(β-pinene) was achieved only for low molecular weights.

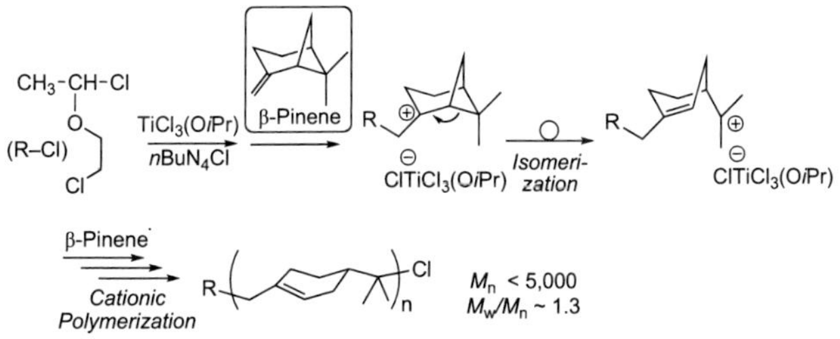

Scheme 3 Living Cationic Polymerizations of β-Pinene

The synthesis of a higher molecular weight poly(terpene) was accomplished by controlling the cationic polymerization that was followed by hydrogenation into bio-based cycloolefin polymers [51]. The cationic polymerizations of β-pinene and α-phellandrene were examined using a series of Lewis acid catalysts at –78°C in various solvents, in which a small amount of adventitious water may act as an initiator (protogen or cationogen). Whereas (–)-α-phellandrene was polymerized by Lewis acids into low molecular weight polymers ($M_w \sim 1 \times 10^4$), relatively high molecular weight polymers ($M_w > 5 \times 10^4$) were obtained from β-pinene using some aluminum-based Lewis acids under judicious polymerization conditions. For further controlling the molecular weight, living cationic polymerization of β-pinene was then investigated in conjunction with $EtAlCl_2$ as the Lewis acid (activator) and the hydrogen chloride adduct of α-methylstyrene as the initiator (cationogen). Although the polymerization with the HCl/$EtAlCl_2$ system was instantaneous even at –78°C in the solvent mixture of CH_2Cl_2/cyclohexane, the number-average molecular weights (M_n) of the produced polymers were close to the calculated values assuming that one initiator molecule generates one living polymer chain. Furthermore, upon the addition of Lewis bases such as diethyl ether and ethyl acetate the polymerization was moderately retarded to afford polymers with relatively narrow molecular weight distributions (MWDs) ($M_w/M_n \sim 1.2$). Much higher molecular weight polymers were produced with the HCl/Al-based system, of which the weight-average molecular weights reached to the high tens of thousands.

In general, the hydrogenation of unsaturated polymers provides a higher thermal resistance and durability to the polymers. The chemical or catalytic hydrogenation of the high molecular weight poly(β-pinene) was then carried out with p-toluenesulfonyl hydrazide in o-xylene or catalytically with Pd (5 wt%)/Al_2O_3 in hexane under 1.0 MPa hydrogen atmosphere (Scheme 4). Especially with the Pd catalyst, the quantitative hydrogenation proceeded to afford a saturated alicyclic hydrocarbon polymer without a significant chain scission (> 99.9%). The thermal properties of the bio-based cycloolefin polymer were then evaluated by DSC and thermogravimetric analyses (TGA). The glass transition temperatures (T_g) were improved by hydrogenation from 90°C to 130°C. Whereas the thermal degradation of the unsaturated poly(β-pinene) proceeded around 300°C, that of the bio-based cycloolefin polymer occurred above 400°C (5% weight loss at 450°C). Thus, the bio-based cycloolefin polymer has both a better thermal durability and higher service temperature than the unsaturated precursor. In addition, the bio-based cycloolefin polymer not only has good thermal properties but also excellent optical properties as well, of which the transparency was almost the same as that of PMMA (visible light transmittance: ~ 92%/1 mm).

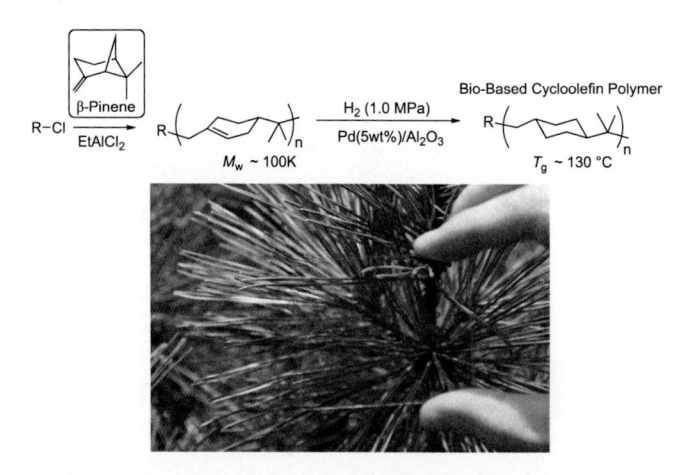

Figure 2   Bio-Based Cycloolefin Polymer from β-Pinene

Controlled radical copolymerization of terpene

Apart from cationic polymerization, monoterpenes are also known to rarely undergo radical homopolymerization except for conjugated dienes like myrcene. Some examples have been reported for their radical copolymerizations with polar monomers, such as maleic anhydride and acrylates, although they resulted in the very low consumption of terpenes along with its low incorporation into copolymers.

β-Pinene is also known to be copolymerized with various petrochemical vinyl monomers to introduce its ring-opened unit, as in the cationic polymerization, into the copolymers via β-scission of the growing radical [52–54]. Lu et al. reported controlled/living radical copolymerizations of β-pinene with several vinyl monomers using the RAFT system (Scheme 4) [55–58]. Acrylonitrile (AN), methyl acrylate (MA), *n*-butyl acrylate (BA), and maleimide derivatives were employed as comonomers for the β-pinene copolymerization, all of which can be homopolymerized via radical polymerization. As with the case of the copolymerization of polar monomers with an olefin, the copolymerization of β-pinene proceeded to afford copolymers with lower β-pinene contents. The RAFT copolymerization of β-pinene with *N*-phenylmaleimide (PhMI) in dichloroethane exhibited the typical features of a controlled manner with better consumption of β-pinene, although severe retardation was observed in the copolymerization with *N*-alkylmaleimides [58].

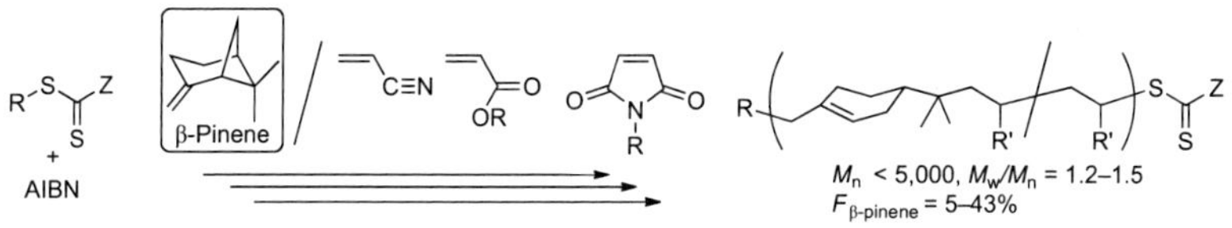

Scheme 4 RAFT Copolymerization of β-Pinene

Among the various terpenes, *d*-limonene (Lim) is one of the most representative popular and abundant compounds, of which the annual production could be estimated to be over a hundred thousand tons per year, and is mainly used in the flavor and fragrance industry [36,37,41]. However, Lim had also been only slightly copolymerized into polymers, while there have been some reports on the free alternating radical copolymerization of Lim with petrochemical vinyl monomers [59–61]. More recently, the authors reported that the quantitative copolymerizations of Lim with maleimide derivatives, such as PhMI and *N*-cyclohexylmaleimide (CyMI), easily occurred in fluorinated cumyl alcohol [PhC(CF$_3$)$_2$OH] to produce not only the chiral and high T_g copolymers with their quantitative conversions but also the unprecedented 1:2 sequence-regulated copolymers from the initiating to the growing end [62]. Lim and maleimide were smoothly consumed with AIBN in PhC(CF$_3$)$_2$OH at an almost 1:2 ratio to produce copolymers with moderate molecular weights, of which the 1:2 cross propagation was proved by their monomer reactivity ratio with a penultimate model. The selective propagation is assumed to be due to the interaction of the fluorinated alcohol as a solvent with the carbonyl groups in the maleimide unit and the bulkiness of the alicyclic structure in Lim. As for the properties of the copolymers obtained from limonene and the maleimides, they exhibited a relatively higher T_g (220–250°C) than those obtained from the usual α-olefins, such as 1-hexene, due to the higher incorporation of maleimides as well as the rigid alicyclic structure of the terpenes [27]. In addition, the copolymers obtained from the chiral limonene showed optical activities and circular dichroisms because the chiral center of the monomer units retained its configuration during the polymerization. Furthermore, the controlled/living radical copolymerization of Lim and PhMI was then examined using *n*-butyl cumyl trithiocarbonate (CBTC) as a RAFT agent with AIBN as the radical reservoir in PhC(CF$_3$)$_2$OH. The control of the molecular weight of the copolymers was demonstrated by the SEC, where the M_n increased with the conversions retaining a unimodal curve. Furthermore, the matrix-assisted laser desorption-ionization time-of-flight mass spectrometry (MALDI-TOF-MS) analysis of the copolymers showed that the copolymers not only consist of the selective 1:2 sequence, but also the well-defined initiating and capping sequence (Figure 3).

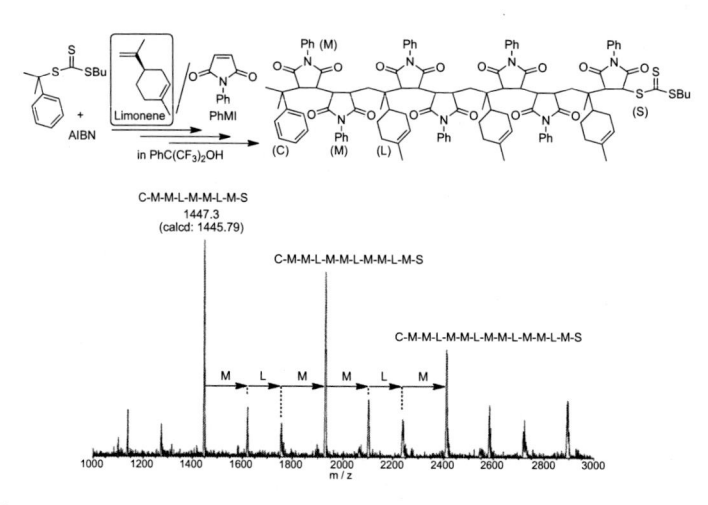

Figure 3 AAB-Sequence Living Radical Copolymerization of d-Limonene.

Other controlled polymerizations of terpenes

As for anionic polymerization of terpene, it had been reported that myrcene can be polymerized into living polymers due to its conjugated diene structure similar to isoprene and butadiene [63]. Quirk et al. reported that the living anionic polymerization with an alkyl lithium as the initiator in non-polar solvents allowed the block copolymerization of myrcene with styrenes even for polystyrene-b-poly(myrcene)-b-polystyrene triblock copolymer, which could be employed as a thermoplastic elastomer consisting of hard polystyrene and soft poly(myrcene) segments (Scheme 5) [64,65].

More recently, Hoye and Hillmyer reported that myrcene can be converted into another cyclic diene, 3-methylenecyclopentene, by ring-closing metathesis (RCM) with the second generation Grubbs-type ruthenium catalyst along with the elimination of isobutene [66]. The cyclic monomer could be polymerized into a cycloolefin polymer by both anionic and cationic polymerization systems. As with anionic polymerizations of other dienes, an alkyl lithium such as sec-butyl lithium leads to the living polymer of 3-methylenecyclopentene with narrow MWDs (M_w/M_n ~ 1.2), and the content of the 4,3-repeating units in the resultant polymer became higher upon the addition of the bidentate Lewis base, N, N, N', N'-tetramethyl ethylenediamine (TMEDA). The living cationic polymerization of 3-methylenecyclopentene was attained by the $HCl/ZnCl_2$ system to afford living polymers not only with controlled molecular weights and narrow MWDs but also with 1,4-regioselective structures, although the polymers were semicrystalline even after hydrogenation.

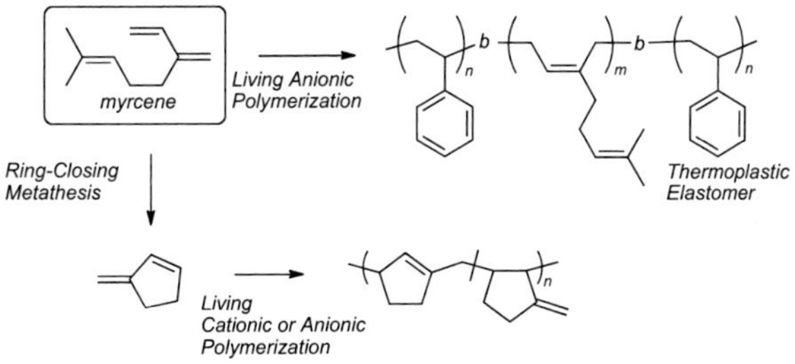

Scheme 5 Living Polymerizations of Myrcene and Its Derivatives

4.1.4 Polymerizations of naturally-occurring styrenes (phenylpropanoids)

Cationic polymerizations of natural β-methylstyrene

In the vinyl-polymerization industry, β-substituted styrene derivatives have rarely been employed as raw materials because such 1,2-disubstituted olefins exhibit little or no tendency to undergo homopolymerizations due to their steric inhibition [9]. On the other hand, such compounds and their polymers (though not via vinyl polymerization) are often found in natural products, which are referred to as *phenylpropanoids* since those contain a phenyl ring with a C_3 side chain derived from phenylalanine and/or tyrosine. Some of them bear β-methylstyrene skeletons that are also abundantly produced in natural plants. For example, *trans*-anethole (Ane: 4-methoxy-β-methylstyrene) is extracted from anise or fennel oil, and isoeugenol (IEu: 4-hydroxy-3-methoxy-β-methylstyrene) or its precursor, eugenol, from clove or ylang-ylang oil, thousands of tons of which are annually used in the flavor and fragrance industry [67]. Although there are several reports on the conventional cationic (co)polymerizations of these β-methylstyrenes, they afforded ill-defined polymers or resulted in low molecular weight oligomers in low yields [68–73].

The authors recently investigated living cationic polymerization of naturally-occurring *trans*-anethole (Ane) with the R–Cl/SnCl$_4$/nBu$_4$NCl initiating system (Scheme 6) [74]. Despite the sterically-hindered β-methyl group, Ane was polymerized in a controlled manner to afford polymers with a controlled molecular weight under the appropriate conditions ([SnCl$_4$]$_0$ < [nBu$_4$NCl]$_0$) [75], whereas the molecular weight distribution of the polymers were relatively broad at a lower concentration of nBu$_4$NCl probably due to the intermolecular Friedel-Crafts reaction. At a lower polymerization temperature, a narrower molecular weight distribution was obtained. Furthermore, the block copolymerization of Ane with petrochemical *p*-methoxylstyrene (pMOS) was also examined using the R–Cl/SnCl$_4$/nBu$_4$NCl system, where M_n of the obtained block copolymers increased while maintaining relatively narrow molecular weight distributions. With the SnCl$_4$-based system, another phenylpropanoid isosafrole (3,4-methylenedioxy β-methylstyrene), which bears a rigid 1,3-dioxolane ring, can also be copolymerized with pMOS in a controlled fashion to afford copolymers with moderate contents of isosafrole and relatively narrow MWDs, although the homopolymerization of isosafrole resulted in low molecular weight oligomers due to the indane ring-forming trimerization via the Friedel-Crafts reaction.

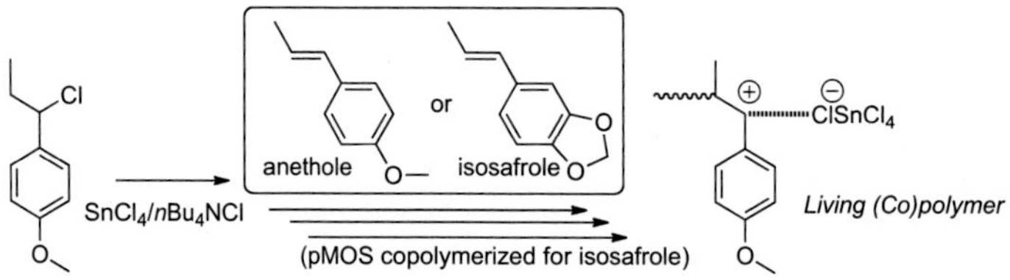

Scheme 6 Living Cationic (Co)polymerization of Naturally-Occurring β-Methylstyrenes

The most abundant natural poly(phenylpropanoid) is lignin, which is the main component of the plant cell and mainly consists of the coniferyl [(4-hydroxy-3-methoxyphenyl)propenyl] moiety that has many biochemical functions, such as an anti-oxidant and sunburn protection due to the phenolic hydroxyl group [76]. However, its complex cross-linked structure, formed by oxidative coupling biosynthesis, has not been realized in synthetic polymers. Among various naturally-occurring β-methylstyrenes, IEu is one of the most promising phenolic monomers as a model for the lignin components in terms of the coniferyl-containing structure, availability, and expected reactivity. Although the recently developed living radical polymerization might be the most suitable for controlling the (co)polymerization of IEu due to the highly tolerant nature to functional groups, the phenolic moiety is fatal to the radical polymerization due to its antioxidative properties. In contrast, similar, but different phenol-based petrochemical-derived vinyl polymers are applicable as photoresists, epoxy-curing agents, and anti-oxidants. The authors have found a unique and rather surprising system to control the cationic polymerization of styrene derivatives including a phenolic monomer, p-hydroxystyrene (pHS), which can be directly polymerized into living polymers without any protection of its phenolic group in the presence of borontrifluoride etherate (BF_3OEt_2) coupled with an alcohol as the initiator [77–79]. In sharp contrast to the conventional living cationic polymerizations in rigorously dried reaction media [11,12], the system is also unique in requiring a fairly large amount of water (approximately equimolar to the monomer and large excess over BF_3OEt_2) for controlling the polymerization.

Although the homopolymerization of IEu did not proceed at all, the copolymerization of IEu with pMOS by the alcohol/BF_3OEt_2 system afforded copolymers with controlled molecular weights and MWDs (Figure 4) [80]. IEu and pMOS were simultaneously consumed by BF_3OEt_2 without the protection of the phenol group in the presence of water, in which the consumption rate of IEu was almost the same as that of pMOS. The M_n's of the producing copolymers increased in direct proportion to the monomer conversions retaining relatively narrow MWDs (M_w/M_n = 1.2–1.4), which were close to the calculated values assuming that one molecule of the alcohol generates one living polymer chain. Interestingly, this copolymerization not only proceeds well but gives almost alternating copolymers as well, which was unprecedented in cationic polymerization. The copolymer composition curve for IEu and pMOS exhibited a typical alternating line, in which both the monomer reactivity ratios are much lower than unity and close to zero, and the MALDI-TOF-MS analysis also exhibited an alternating fashion, where the highest series of peaks are alternately separated by the molecular weight of the pMOS and IEu monomers. This alternating copolymerization would be most probably due to the highly bulky and electron-rich IEu. The growing cation of the sterically hindered IEu would not react with the bulky monomer, whereas the less hindered styryl cation favors the more electron-rich IEu.

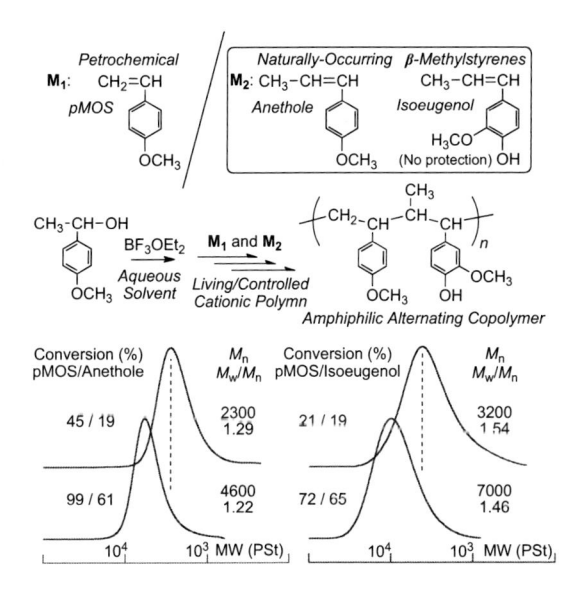

Figure 4 Phenolic Alternating Copolymer by Controlled Cationic Copolymerization of Isoeugenol.

Controlled radical copolymerizations of phenylpropanoids.

Similar to the terpenes described above, the 1,2-disubstituted olefin of phenylpropanoid has sometimes been employed as a non-homopolymerizable monomer in the copolymerization with petrochemical-derived monomers [81,82]. As examples, Ane has been used as an electron-donating monomer for the copolymerization with electron-withdrawing monomers, such as maleic anhydride, AN, and acrylates. Whereas the Ane copolymerization with homopolymerizable AN or (meth)acrylates produced copolymers with low Ane contents, the combination with non-homopolymerizable maleic anhydride exhibited a highly alternating tendency [83]. The controlled/living radical copolymerization of Ane or isosafrole was examined with common vinyl monomers, such as MA, methyl methacrylate, and styrene, where 2-cyanoprop-2-yl ethyl trithiocarbonate (CPETC) was employed as a RAFT agent in conjunction with AIBN as the radical source (Scheme 7) [84]. Especially for the copolymerization of Ane and MA, the M_n of the obtained copolymers increased in direct proportion to the monomer conversions and were close to the calculated values assuming that one RAFT agent generates one polymer chain, and the MWDs were relatively narrow during the polymerization ($M_w/M_n \sim 1.2$). Furthermore, the Ane incorporation into the copolymers depended on the reaction solvents. Among the various solvents, a fluorinated alcohol m-$C_6H_4[C(CF_3)_2OH]_2$ gave the highest Ane content (Ane/MA = 41/59), indicating that the fluoroalcohol as a solvent enhanced the copolymerizability and led to the higher incorporation of Ane into the produced copolymers via a hydrogen-bonding interaction with the carbonyl moieties of the acrylic monomer units as in the copolymerization of α-olefin [85]. Especially, when the copolymerization was conducted in a fluoroalcohol at a higher Ane feed ratio ($[Ane]_0/[MA]_0=9/1$), the obtained copolymers showed an equimolar content of Ane and MA, indicating the formation of almost alternating copolymers, which was analyzed by MALDI-TOF-MS for determining the terminal and sequence structures of the copolymers. The copolymerization of isosafrole with MA in m-$C_6H_4[C(CF_3)_2OH]_2$ also proceeded in a living fashion to give almost alternating copolymers with relatively narrow MWDs along with a high T_g due to the rigid 1,3-dioxolane ring.

Scheme 7 RAFT Copolymerization of Naturally-Occurring β-Methylstyrenes

Cinnamic acid derivatives are other non-homopolymerizable phenylpropanoids that can also be abundantly found in natural products like cassia and cinnamon oil [82,86,87], although most of them have been commercially produced via the aldol reactions of benzaldehyde originating from petrochemicals. These compounds seem interesting because they bear not only the styrene skeleton but also an acrylic character. The radical copolymerization of methyl cinnamate (MC) was examined with common vinyl monomers, such as MA and styrene. MC was readily copolymerized with both MA and styrene in various solvents to give copolymers with relatively high molecular weights, although the incorporation ratio of MC into the copolymers was independent of the solvent (30–40%). Furthermore, the living radical copolymerizations of MC and the vinyl monomers were examined by the RAFT copolymerization using CPETC in toluene and m-$C_6H_4[C(CF_3)_2OH]_2$, or by the transition-metal catalyzed atom transfer radical polymerization using RuCp*Cl(PPh$_3$)$_2$ coupled with (MMA)$_2$-Cl as the initiator in toluene (Scheme 8) [88]. In both cases, the M_n of the produced copolymers increased in direct proportion to the monomer conversions and were close to

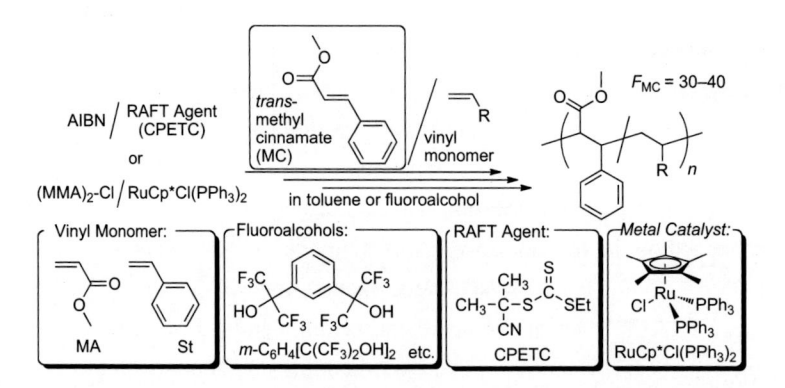

Scheme 8 Living Radical Copolymerizations of Cinnamates

the calculated values, and the molecular weight distributions (MWDs) were narrower throughout the polymerization.

4. 1. 5 Naturally-derived acrylic monomers

Tulipalin A (α-methylene butyrolactone) is a vinyl monomer consisting of a five-membered ring with an exo-methylene moiety adjacent to a carbonyl group, which is a natural product found in tulips [2]. The monomer and its derivatives were reported to be homopolymerized in similar manners to methacrylates such as radical, anionic, group transfer, and coordination polymerizations, although the monomers examined in the polymerizations were originally derived from petrochemical products (Scheme 9) [89–92]. The free radical polymerizations of tulipalin A produced polymers with a higher service temperature ($T_g > 190°C$) than that from methyl methacrylate (~ 110°C) due to the rigid lactone ring, which also provides the polymer with a good durability and a good refractive index. The controlled radical polymerization of tulipalin A with the Cu(I)-catalyzed ATRP was reported by Matyjaszewski et al [93,94]. Tulipalin A was polymerized with an initiating system consisting of R-Br/CuCl/CuCl$_2$/bipyridine to produce the polymers with relatively narrow MWDs ($M_w/M_n < 1.2$), and the block copolymerization with butyl acrylates could be a thermoplastic elastomer, in which the poly(tulipalin A) segment acts as the hard and binding phase. α-Methylene-γ-methyl-γ-butyrolactone has an exo-methylene five-membered lactone similar to tulipalin A, which can be prepared from the biomass-derived levulinic acid [95]. The γ-methyl version also has a homopolymerizability via various polymerization methods to give high T_g polymers, of which the controlled radical polymerization was achieved using the RAFT system in a heterogeneous miniemulsion system [96]. Chen et al. recently reported a fast and living group transfer polymerization system for these renewable methylene butyrolactones consisting of a silyl ketene acetal as the initiator and trityl borate as the activator [97]. The polymerizations with triisobutylsilyl ketene acetal combined with Ph$_3$CB(C$_6$F$_5$)$_4$ proceeded fast in a living fashion to give relatively high molecular weight polymers with very narrow MWDs ($M_n > 500K$, $M_w/M_n = 1.01$). They also reported a living coordination polymerization system with early transition-metal catalysts [92].

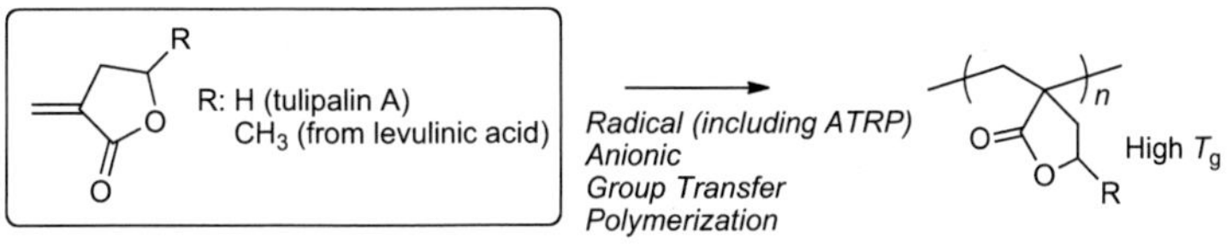

Scheme 9 Polymerizations of α-Methylene Butyrolactones

Another candidate for bio-based acrylic monomers should be the derivatives of itaconic acid (methylenesuccinic acid), which has been produced via fermentation of the carbohydrate like lactic acid or pyrolysis of natural citric acid, although they are conventional and widely-used vinyl monomers. The polymerization of itaconic acid and its derivatives had been reported since the 1870s via radical and anionic polymerizations to produce relatively high molecular weight polymers, of which the reactivity and the molecular weights of the polymer depend on their substituents [98–104]. Furthermore, the diamide and imide derivatives of itaconic acid could also be polymerized via anionic and radical intermediates, among which the substituted itacoimides gave excellent thermal properties ($T_g > 220°C$) due to their cyclic structures [105,106]. Recently, the controlled/living radical polymerizations of itaconic acid derivatives were also investigated. The RAFT polymerization of alkyl itaconates gave polymers with controlled molecular weights though the MWDs broadened [107,108]. The Cu(I)-catalyzed ATRP was employed for the block copolymerization of N-aryl itaconimide with MMA [109]. The authors also examined the controlled/living radical polymerization of itaconic acid derivatives, such as N-phenylitaconimide (NPI) and di-n-butylitaconate (DBI), with a RAFT agent, in which the M_n of the obtained polymer increased with conversions retaining relatively narrow MWDs and agreed with the calculated values (Scheme 10) [110].

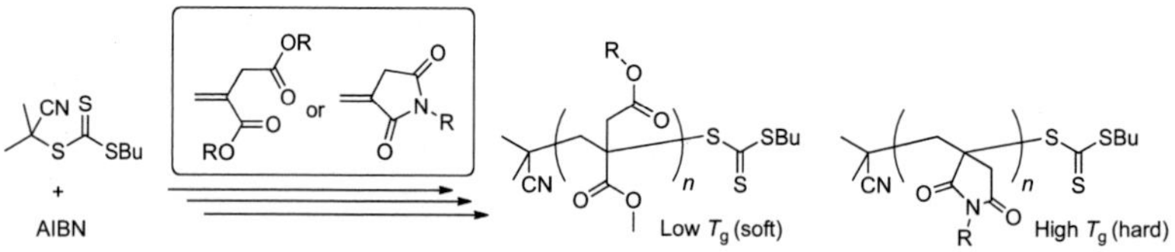

Scheme 10 RAFT Polymerization of Itaconic Acid Derivatives

4. 1. 6 Conclusion

Using the technology developed for conventional polymer synthesis from petrochemicals, the unprecedented bio-based polymers with high-performance or functions have been synthesized from naturally-occurring or -derived vinyl monomers. However, in order to create environmentally benign materials meeting social needs and to establish a new industry, it will be required to satisfy the market needs with neither losing any performance as materials nor costing more. The developments in polymerizations would contribute to further progress in bio-based polymeric materials from renewable vinyl monomers leading to commercial production in the near future.

References

1. D. L. Klass, Biomass for Renewable Energy, Fuels, and Chemicals, Academic Press, San Diego (1998)
2. R. Mullin, *Chem. Eng. News*, **82**(45), 29–37 (2004)
3. R. P. Wool, X. S. Sun, Bio-Based Polymers and Composites, Elsevier, Oxford (2005)
4. N. N. Belgacem, A. Gandini, Monomers, Polymers and Composites from Renewable Resources, Elsevier, Oxford (2008)

5. A. Gandini, *Macromolecules*, **41**(24), 9491–9504 (2008)
6. C. K. Williams, M. A. Hillmyer, *Polymer Reviews*, **48**(1), 1–10 (2008)
7. Y. Kimura, *Polym. J.*, **41**(10), 797–807 (2009)
8. G. W. Coates, M. A. Hillmyer, *Macromolecules*, **42**(21), 7987–7989 (2009)
9. G. Odian, Principles of Polymerization, 4th ed., John Wily and Sons, Inc., Hoboken, NJ (2004)
10. M. Szwarc, *Nature*, **178**(4543), 1168–1169 (1956)
11. M. Sawamoto, *Prog. Polym. Sci.*, **16**(1), 111–172 (1991)
12. S. Aoshima, S. Kanaoka, *Chem. Rev.*, **109**(11), 5245–5287 (2009)
13. M. K. Georges, R. P. N. Veregin, P. M. Kazmaier, G. K. Hamer, *Macromolecules*, **26**(11), 2987–2988 (1993)
14. C. J. Hawker, *J. Am. Chem. Soc.*, **116**(24), 11185–11186 (1994)
15. C. J. Hawker, A. W. Bosman, E. Harth, *Chem. Rev.*, **101**(12), 3661–3688 (2001)
16. M. Kato, M. Kamigaito, M. Sawamoto, T. Higashimura, *Macromolecules*, **28**(5), 1721–1723 (1995)
17. J. S. Wang, K. Matyjaszewski, *J. Am. Chem. Soc.*, **117**(20), 5614–5615 (1995)
18. V. Percec, B. Barboiu, *Macromolecules*, **28**(23), 7970–7972 (1995)
19. C. Granel, P. Dubois, R. Jerome, P. Teyssie, *Macromolecules*, **29**(27), 8576–8582 (1996)
20. D. M. Haddleton, C. B. Jasieczek, M. J. Hannon, A. J. Shooter, *Macromolecules*, **30**(7), 2190–2193 (1997)
21. K. Matyjaszewski, J. H. Xia, *Chem. Rev.*, **101**(9), 2921–2990 (2001)
22. M. Kamigaito, T. Ando, M. Sawamoto, *Chem. Rev.*, **101**(12), 3689–3745 (2001)
23. J. Chiefari, Y. K. Chong, F. Ercole, J. Krstina, J. Jeffery, T. P. T. Le, R. T. A. Mayadunne, G. F. Meijs, C. L. Moad, G. Moad, E. Rizzardo, S. H. Thang, *Macromolecules*, **31**(16), 5559–5562 (1998)
24. D. Charmot, P. Corpart, H. Adam, S. Z. Zard, T. Biadatti, G. Bouhadir, *Macromol. Symp.*, **150**, 23–32 (2000)
25. M. Destarac, D. Charmot, X. Franck, S. Z. Zard, *Macromol. Rapid Commun.*, **21**(15), 1035–1039 (2000)
26. G. Moad, E. Rizzardo, S. H. Thang, *Aust. J. Chem.*, **58**(6), 379–410 (2005)
27. V. Dragutan, R. Streck, Catalytic Polymerization of Cycloolefins, Studies in Surface Science and Catalysis, vol. 131, Elsevier Science, Amsterdam (2000)
28. G. W. Coates, *Chem. Rev.*, **100**(4), 1223–1252 (2000)
29. T. M. Trnka, R. H. Grubbs, *Acc. Chem. Res.*, **34**(1), 18–29 (2001)
30. I. Tritto, L. Boggioni, D. R. Ferro, *Coord. Chem. Rev.*, **250**(1–2), 212–241 (2006)
31. H. Cherdron, M. J. Brekner, F. Osan, *Angew. Makromol. Chem.*, **223**, 121–133 (1994)
32. T. Inoue, O. Takiguchi, K. Osaki, T. Kohara, T. Natsuume, *Polym. J.*, **26**(2), 133–139 (1994)
33. T. Inoue, H. Okamoto, K. Osaki, T. Kohara, T. Natsuume, *Polym. J.*, **27**(9), 943–950 (1995)
34. J. Kodemura, T. Natsuume, *Polym. J.*, **27**(12), 1167–1172 (1995)
35. T. Otsuki, K. Goto, Z. Komiya, *J. Polym. Sci., Part A, Polym. Chem.*, **38**, 4661–4668 (2000)
36. R. J. Braddock, Handbook of Citrus By-Products and Processing Technology, John Wiley & Sons, New York (1999)
37. E. Breitmaier, Terpenes, Wiley-VCH, Weinheim (2006)
38. W. J. Roberts, A. R. Day, *J. Am. Chem. Soc.*, **72**(3), 1226–1230 (1950)
39. Kennedy, J. P., Cationic Polymerization of Olefins, A Critical Inventory, Wiley-Interscience, New York (1975)
40. J. P. Kennedy, E. Maréchal, Carbocationic Polymerization, John Wiley and Sons, Inc., Hoboken, NJ (1982)
41. J. J. W. Coppen, G. A. Hone, Gum Naval Stores, Turpentine and Rosin from Pine Resin, Natural Resource Institute, Food and Agriculture Organization of the United Nations, Rome (1995)

42. J. H. Langenheim, Plant Resins, Timber Press Inc, Portland, OR (2003)
43. B. Keszler, J. P. Kennedy, *Adv. Polym. Sci.*, **100**, 1–9 (1992)
44. J. Lu, M. Kamigaito, M. Sawamoto, T. Higashimura, Y. X. Deng, *Macromolecules*, **30**(1), 22–26 (1997)
45. J. Lu, M. Kamigaito, M. Sawamoto, T. Higashimura, Y. X. Deng, *Macromolecules*, **30**(1), 27–31 (1997)
46. J. Lu, M. Kamigaito, M. Sawamoto, T. Higashimura, Y. X. Deng, *J. Polym. Sci., Part A, Polym. Chem.*, **35**(8), 1423–1430 (1997)
47. L. Hui, L. Jiang, *J. Appl. Polym. Sci.*, **75**(5), 599–603 (2000)
48. J. Lu, H. Liang, R. F. Zhang, B. Li, *Polymer*, **42**(10), 4549–4553 (2001)
49. J. Lu, H. Liang, W. Zhang, Q. Cheng, *J. Polym. Sci., Part A, Polym. Chem.*, **41**(9), 1237–1242 (2003)
50. J. Lu, H. Liang, A. L. Li, Q. Cheng, *Eur. Polym. J.*, **40**(2), 397–402 (2004)
51. K. Satoh, H. Sugiyama, M. Kamigaito, *Green Chemistry*, **8**(10), 878–882 (2006)
52. J. Maslinskasolich, I. Rudnicka, *Eur. Polym. J.*, **24**(5), 453–456 (1988)
53. A. M. Ramos, L. S. Lobo, *Macromol. Symp.*, **127**, 43–50 (1998)
54. M. Paz-Pazos, C. Pugh, *J. Polym. Sci., Part A, Polym. Chem.*, **44**(9), 3114–3124 (2006)
55. A. L. Li, Y. Wang, H. Liang, J. Lu, *J. Polym. Sci., Part A, Polym. Chem.*, **44**(8), 2376–2387 (2006)
56. Y. Wang, A. L. Li, H. Liang, J. Lu, *Eur. Polym. J.*, **42**(10), 2695–2702 (2006)
57. A. L. Li, X. Y. Wang, H. Liang, J. Lu, *React. Funct. Polym.*, **67**(5), 481–488 (2007)
58. Y. Wang, Q. Chen, H. Liang, J. Lu, *Polym. Int.*, **56**(12), 1514–1520 (2007)
59. T. Doiuchi, H. Yamaguchi, Y. Minoura, *Eur. Polym. J.*, **17**(9), 961–968 (1981)
60. J. Maslinskasolich, T. Kupka, M. Kluczka, A. Solich, *Macromol. Chem. Phys.*, **195**(5), 1843–1850 (1994)
61. S. Sharma, A. K. Srivastava, *Eur. Polym. J.*, **40**(9), 2235–2240 (2004)
62. K. Satoh, M. Matsuda, K. Nagai, M. Kamigaito, *J. Am. Chem. Soc.*, **132**(29), 10003–10005 (2010)
63. A. Sivola, *Acta Polytech. Scand., Chem. Technol. Ser.*, **134**, 7–65 (1977)
64. R. P. Quirk, T.-L. Huang, New Monomers and Polymers, Plenum Press, New York, p 329 (1984)
65. R. P. Quirk, Thermoplastic Elastomers, 3rd ed., Hanser, Munich, p 69–91 (2004)
66. S. Kobayashi, C. Lu, T. R. Hoye, M. A. Hillmyer, *J. Am. Chem. Soc.*, **131**(23), 7960–7963 (2009)
67. S. Salvin, B. M., T. Byrne, Eds., The New Crop Industries Handbook, Rural Industries Research and Development Corporation, Canberra (2004)
68. T. Alfrey, L. Arond, C. G. Overberger, *J. Polym. Sci.*, **4**(4), 539–541 (1949)
69. C. G. Overberger, D. Tanner, E. M. Pearce, *J. Am. Chem. Soc.*, **80**(17), 4566–4568 (1958)
70. A. Mizote, T. Tanaka, T. Higashim, S. Okamura, *J. Polym. Sci., Part A*, **3**(7PA), 2567-& (1965)
71. T. Higashim, T. Masuda, K. Kawamura, *J. Polym. Sci., Part A-1, Polym. Chem.*, **10**(1), 85–93 (1972)
72. R. Alexander, A. Jefferson, P. D. Lester, *J. Polym. Sci., Part A, Polym. Chem.*, **19**(3), 695–706 (1981)
73. D. L. Trumbo, *Polym. Bull.*, **33**(6), 643–649 (1994)
74. S. Saitoh, K. Satoh, M. Kamigaito, *Polym. Prepr., Jpn.*, **57**(2), 2965–2966 (2008)
75. T. Higashimura, Y. Ishihama, M. Sawamoto, *Macromolecules*, **26**(4), 744–751 (1993)
76. N. G. Lewis, S. Sarkanen, Lignin and Lignan Biosynthesis (ACS Symposium Series 697), American Chemical Society, Washington, DC (1998)
77. K. Satoh, M. Kamigaito, M. Sawamoto, *Macromolecules*, **33**(15), 5405–5410 (2000)
78. K. Satoh, M. Kamigaito, M. Sawamoto, *Macromolecules*, **33**(16), 5830–5835 (2000)
79. K. Satoh, J. Nakashima, M. Kamigaito, M. Sawamoto, *Macromolecules*, **34**(3), 396–401 (2001)
80. K. Satoh, S. Saitoh, M. Kamigaito, *J. Am. Chem. Soc.*, **129**(31), 9586–9587 (2007)
81. D. Braun, M. Schacht, H. Elsasser, F. Tudos, *Macromol. Rapid Commun.*, **18**(4), 335–342 (1997)
82. D. Braun, F. C. Hu, *Prog. Polym. Sci.*, **31**(3), 239–276 (2006)

83. T. Kokubo, S. Iwatsuki, Y. Yamashita, *Macromolecules*, **3**(5), 518–523 (1970)
84. Y. Nonoyama, K. Satoh, M. Kamigaito, *Polym. Prepr. Jpn.*, **58**(1), 522 (2008)
85. K. Koumura, K. Satoh, M. Kamigaito, *Macromolecules*, **42**(7), 2497–2504 (2009)
86. C. A. Barson, *J. Polym. Sci.*, **62**(174), S128-& (1962)
87. K. Fujimori, W. S. Schiller, I. E. Craven, *Makromol. Chem.-Macromol. Chem. Phys.*, **192**(4), 959–966 (1991)
88. Y. Terao, K. Nagai, K. Satoh, M. Kamigaito, *Polym. Prepr. Jpn.*, **59**(1), 490 (2010)
89. M. K. Akkapeddi, *Macromolecules*, **12**(4), 546–551 (1979)
90. M. K. Akkapeddi, *Polymer*, **20**(10), 1215–1216 (1979)
91. J. Suenaga, D. M. Sutherlin, J. K. Stille, *Macromolecules*, **17**(12), 2913–2916 (1984)
92. G. M. Miyake, S. E. Newton, W. R. Mariott, E. Y. X. Chen, *Dalton Trans.*, **39**(29), 6710–6718 (2010)
93. J. Mosnacek, J. A. Yoon, A. Juhari, K. Koynov, K. Matyjaszewski, *Polymer*, **50**(9), 2087–2094 (2009)
94. J. Mosnacek, K. Matyjaszewski, *Macromolecules*, **41**(15), 5509–5511 (2008)
95. L. E. Manzer, *ACS Symp. Ser.*, **921**, 40–51 (2006)
96. G. G. Qi, M. Nolan, F. J. Schork, C. W. Jones, *J. Polym. Sci., Part A, Polym. Chem.*, **46**(17), 5929–5944 (2008)
97. G. M. Miyake, Y. T. Zhang, E. Y. X. Chen, *Macromolecules*, **43**(11), 4902–4908 (2010)
98. B. E. Tate, *Adv. Polym. Sci.*, **5**, 214–232 (1967)
99. C. S. Marvel, T. H. Shepherd, *J. Org. Chem.*, **24**(5), 599–605 (1959)
100. S. Nagai, *Bull. Chem. Soc. Jpn.*, **36**(11), 1459–1463 (1963)
101. S. Ishida, S. Saito, *J. Polym. Sci., Part A-1, Polym. Chem.*, **5**(4PA1), 689-& (1967)
102. T. Sato, S. Inui, H. Tanaka, T. Ota, M. Kamachi, K. Tanaka, *J. Polym. Sci., Part A, Polym. Chem.*, **25**(2), 637–652 (1987)
103. T. Otsu, J. Z. Yang, *Polym. Int.*, **25**(4), 245–251 (1991)
104. T. Hirano, S. Tateiwa, M. Seno, T. Sato, *J. Polym. Sci., Part A, Polym. Chem.*, **38**(13), 2487–2491 (2000)
105. H. Watanabe, A. Matsumoto, T. Otsu, *J. Polym. Sci., Part A, Polym. Chem.*, **32**(11), 2073–2083 (1994)
106. H. Watanabe, A. Matsumoto, T. Otsu, *J. Polym. Sci., Part A, Polym. Chem.*, **32**(11), 2085–2091 (1994)
107. Z. Szablan, A. A. Toy, T. P. Davis, X. J. Hao, M. H. Stenzel, C. Barner-Kowollik, *J. Polym. Sci., Part A, Polym. Chem.*, **42**(10), 2432–2443 (2004)
108. Z. Szablan, A. A. Toy, A. Terrenoire, T. P. Davis, M. H. Stenzel, A. H. E. Muller, C. Barner-Kowollik, *J. Polym. Sci., Part A, Polym. Chem.*, **44**(11), 3692–3710 (2006)
109. X. H. Liu, G. B. Zhang, B. X. Li, Y. G. Bai, D. Pan, Y. S. Li, *Eur. Polym. J.*, **44**(4), 1200–1208 (2008)
110. D.-H. Lee, K. Nagai, K. Satoh, M. Kamigaito, *Polym. Prepr. Jpn.*, **58**(1), 491 (2010)

4. 2 Biobased polyamides
Sei-ichi Aiba

4. 2. 1 Introduction

Polyamides (PAs), one of engineering plastics, have superior properties such as high tensile strength, high thermostability, high chemical resistance, etc. Hydrogen bonds between amide groups in PA backbone cause these high performance and then PAs are used in automobile, electric appliance, and fiber. For instance melting temperature (Tm) of PAs are in the range from 170 (PA12) to 290 (PA46)°C. Tm changes along with the number of carbon atoms in the repeating units of PA backbone; Number of carbon atom decreasing, Tm increasing. In addition tensile strength of PAs is almost proportional to Tm (Figure 1). In the world about two million t/y of PA resins are produced. There are a wide variety of polyamides such as PA6, PA66, PA610, etc in the market. The first figure of the abbreviation of PA is the number of carbon atoms in a diamine component and next figure is the number of carbon atoms in a diacid component. In the case of single figure this means the number of carbon atoms in ω-amino acid. Main products are PA6 and PA66. Estimated production capacity of PA6 and PA66 in the world in 2009 is about 1 840 000 and 1 375 000 t/y, respectively [1]. World wide demand of PAs was about 2 200 000 t/y but decreased to about 1 800 000 t/y because of economic crisis in 2008; especially the demand in USA and EU decreased clearly but the demand in Asian countries increased and will increase every year. Main application of PA66 is automobile parts and PA6 are used in the fields of automobile and fiber/film. Other PAs such as PA11, PA12, PA46, PA610, and PA612 have

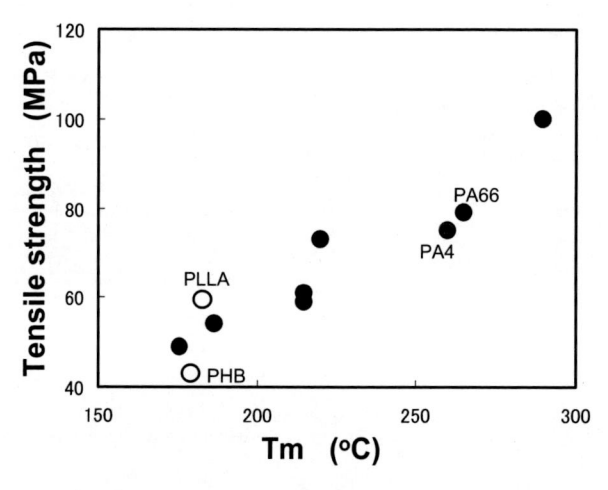

Figure 1 Relationship between tensile strength and melting point;
●, polyamides; ○, commercicalized biobased plastics

4.2.2 Biobased PAs in the market and under R & D

Biobased polymers mean that polymers must be produced from biomass from the viewpoint of carbon source. Usually sucrose, cellulose and starch are used as carbon source. Chemical (catalytic) or biochemical (enzymatic) processes are adopted for their production, for instance poly(3-hydroxy butyric acid) is produced directly from sugar or vegetable oil by microbes. In some cases both processes are used for production of final polymers such as poly(L-lactic acid); lactic acid fermentation by microbes and polymerization of lactic acid by chemical process (Figure 2).

Among the commercialized PAs, PA11 (Arkema, Toyobo) and PA1010 (Evonik Degussa, DuPont) are fully biobased and PA610 (BASF, Toray, DuPont) is partly biobased. PA11 is produced from castor oil via many chemical steps (Figure 3). Main component of castor oil is triglyceride of ricinolic acid. The triglyceride is decomposed by methanol to produce methyl ricinolate and finally 11-amino undecanoic acid is polymerized to PA11. In the case of PA610 sebacic acid is produced from castor oil and then polymerized with 1,6-diamino hexane which is petro-based (Figure 4). The biobased degree is about 60%. PA1010 is produced by polymerization of sebacic acid and 1,10-diamino decane which is also produced from sebacic acid. The biobased degree is 100%. In these case 11-amino undecanoic acid and sebacic acid have advantage over corresponding components produced from petroleum from the economic viewpoint. However production of other monomers from biomass for PAs such as adipic acid, 1,6-diamino hexane, ε-caprolactam, 1,4-amino butane, and so on has less advantage because production of these components from petro-based chemicals has been already established and processes are economically sophisticated. Figure 5 shows the processes for production of these petro-based components. The monomers of PA66 are 1,6-diamino hexane and adipic acid which are produced from propylene via acrylonitrile and adiponitrile. Or adipic acid is produced from benzene via cyclohexene. The monomer of PA6 is

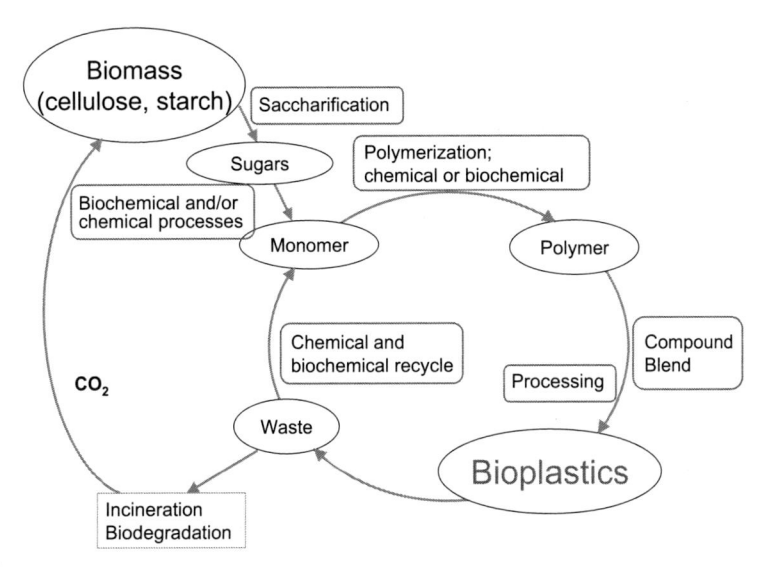

Figure 2 Scheme of cyclic production of bioplastics from biomass

ε-caprolactam which is produced from benzene in petroleum via cyclohexane, cyclohexanone, and cyclohexanone oxim. One of the monomers of PA46 is 1,4-diamino butane which is produced from acrylonitrile.

However, recently much attention have been paid on the production processes for biobased components such as ε-caprolactam [2], 1,4-diamino butane [3], adipic acid [4], and 1,6-diamino hexane. Some patents and literatures reported the possible way to produce these components from biomass. Figure 6 shows the possible processes for production of biobased components. Lysine is commercially produced by fermentation from biomass, the produced amount is next to sodium

Castor oil
↓ Methanolysis
Methyl ricinolate
↓ Pyrolysis
Methyl undecenoate
↓ Hydrolysis
Undecenoic acid
↓ HBr
11-Bromo undecanoic acid
↓ NH$_3$
11-Amino undecanoic acid
↓ Polymerization
PA11

Figure 3   Production process of PA11

Castor oil
↓ Hydrolysis (NaOH)
Ricinoleic acid
↓ Oxidative hydrolysis
Sebacic acid
↓ Polymerization with C6 diamine
PA610

Figure 4   Production process of PA610

Propylene
↓ NH$_3$
Acrylonitrile
↓
Adiponitrile   HCN
↓   ←——— Butadiene
1,6-Diamino hexane and adipic acid
↓
PA66

Benzene
↓
Cyclohexene
←———

Benzene
↓
Cyclohexanone
↓
Cyclohexanone oxime
↓
ε-caprolactam
↓
PA6

Acrylonitrile
↓ HCN
Succinonitrile
↓ H$_2$
1,4-diaminobutane
↓
PA46

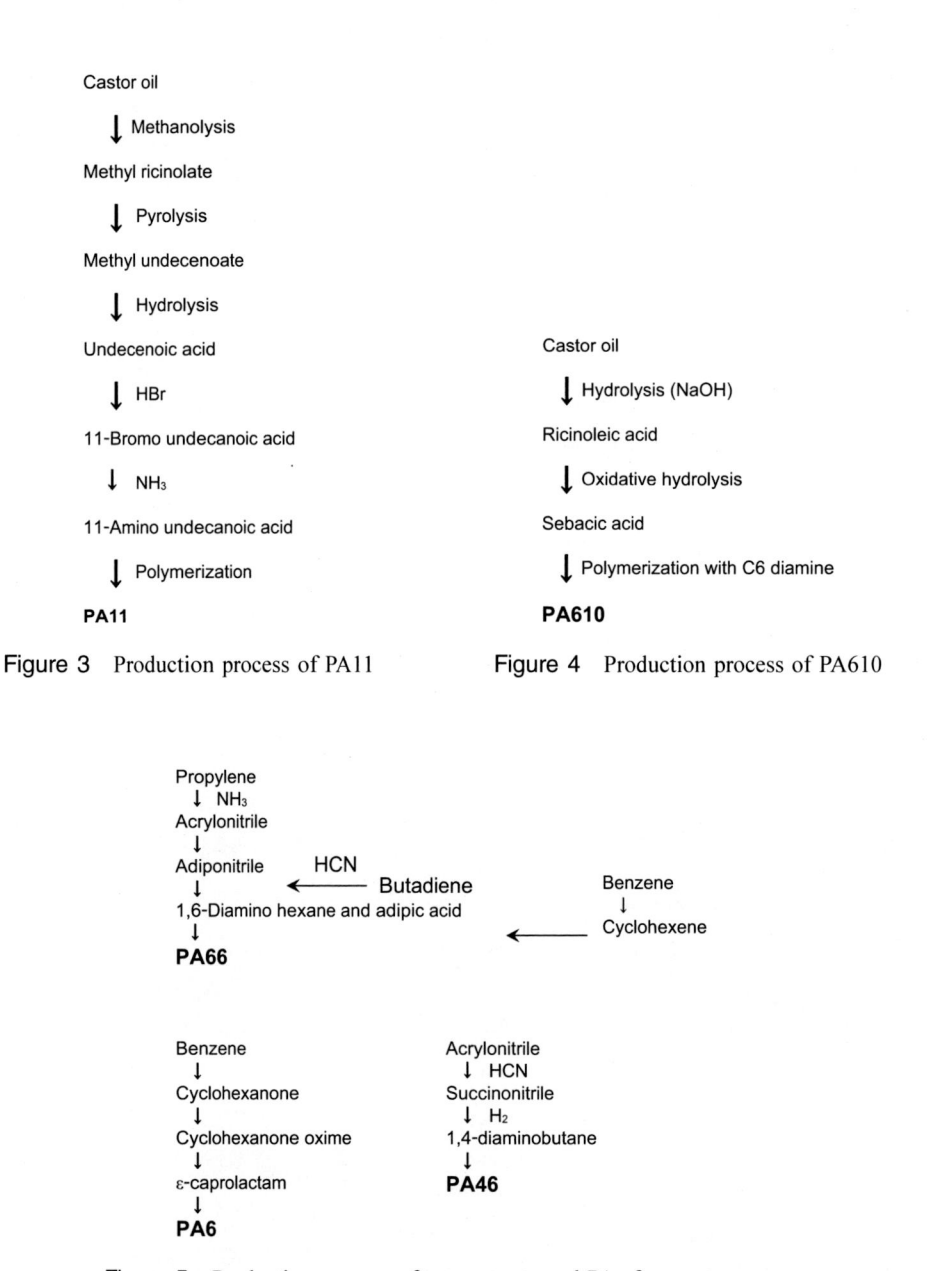

Figure 5   Production process of components and PAs from petroreum

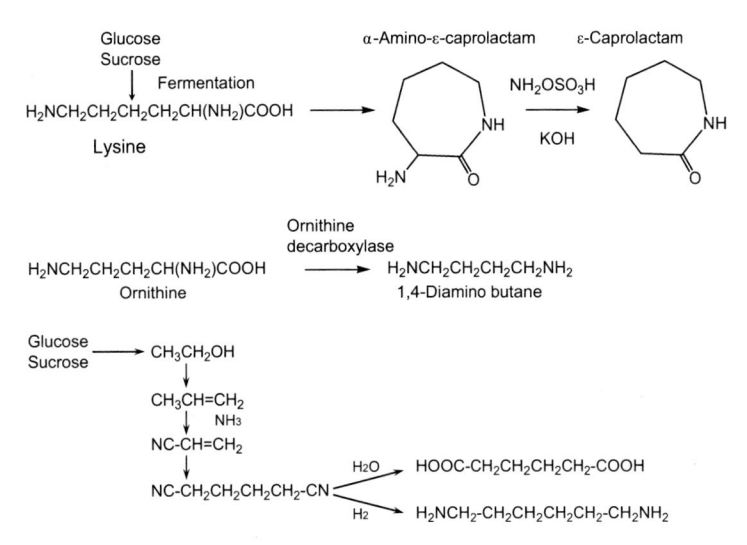

Figure 6 Possible processes for production of conventional components from biomass

glutamate. Lysine was heated in alcohol to produce α-amino-ε-caprolactam and then converted to ε-caprolactam. finally ε-caprolactam was purified by sublimation with total yield of 60–75%. In the case of 1,4-diamino butane, ornithine decarboxylase was effectively applied to ornithine to remove amino group. Maximum yield of 1,4-diamino butane was 5.1 g/L.

Recently bioethanol production is very popular and has commercially advantages, then production of biobased components for polymers from bioethanol has much attention. Dow-Crystalsev and Braskem released the news about commercial production of LLDPE (350 000 t/y) and HDPE (200 000 t/y) through dehydration of ethanol by catalytic processes, respectively. The pass way from ethanol to propylene is also possible, then the production of adipic acid and 1,6-diamino hexane through propylene will receive highlight as shown in Figure 6. Direct fermentative production of adipic acid has no economic advantage now. The biobased PA66 will appear in near future.

The alternative way to produce PAs from biomass is development of novel biobased components which has not been commercialized. In 2004 Department of Energy of USA released the report of "Top Value Added Chemicals from Biomass" [5]. This report nominated 12 building blocks such as glutamic acid, succinic acid, and so on. These are starting chemicals for designing biobased monomers. 2,5-Furandicarboxylic acid (FDCA), one of the 12 building blocks, is seems to be used directly to polymerization. Several papers reported production pass ways for FDCA; oxidation of galactose with nitric followed by heating with p-toluene sulfonic acid [6], synthesis from fructose via hydroxymethyl furfural using cobalt acetylacetonate catalyst with yield of ca.70% [7, 8] (Figure 7), synthesis of dibutyl ester of FDCA from galactaric acid and 1-butanol [9]. Dibutyl furandicarboxylate is useful for next direct polycondensation. Gandini reported the syntheses of PAs using FDCA (Figure 8). PA of Figure 8A has similar chemical structure to Kevlar and high grass transition temperature (Tg), 325 °C [10]. In the case of Figure 8B Tm is 220 °C and Tg 100 °C [11]. Succinic acid is also one of the 12 building blocks and recently commercial production

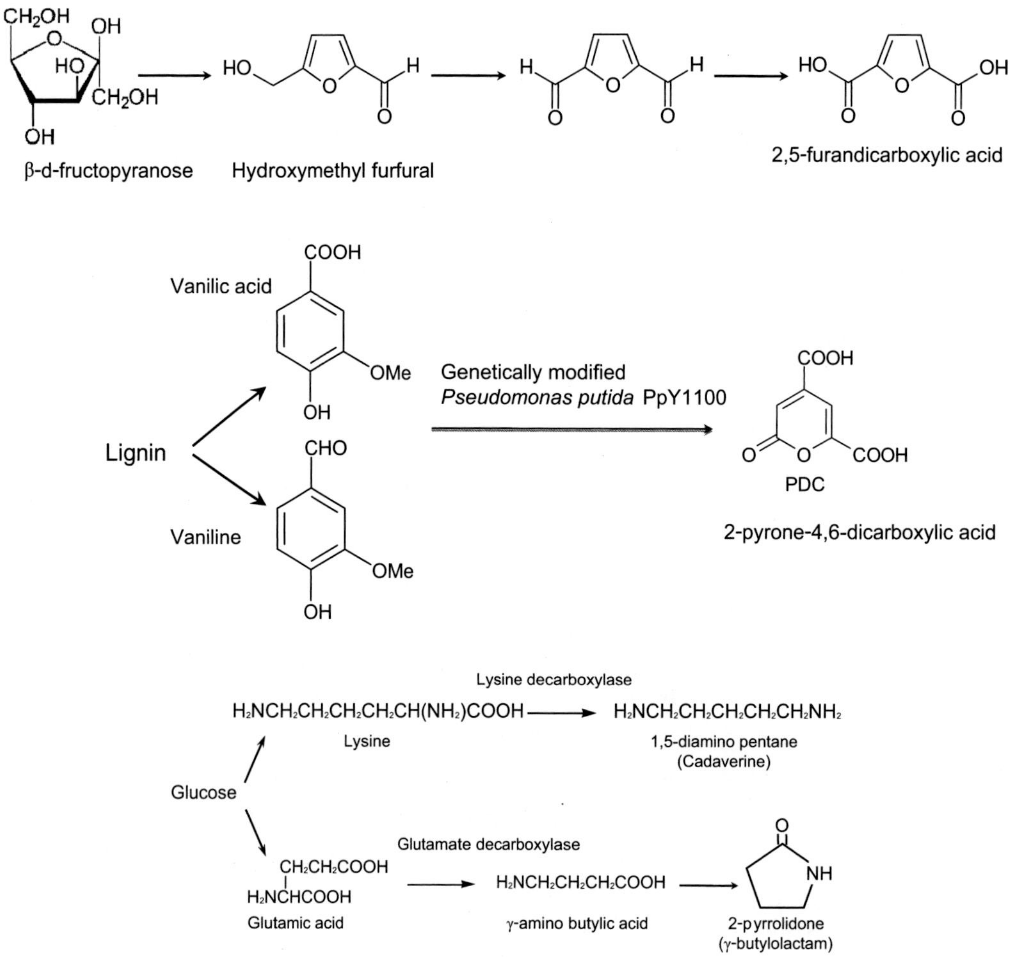

Figure 7 Possible processes for production of novel biobased components

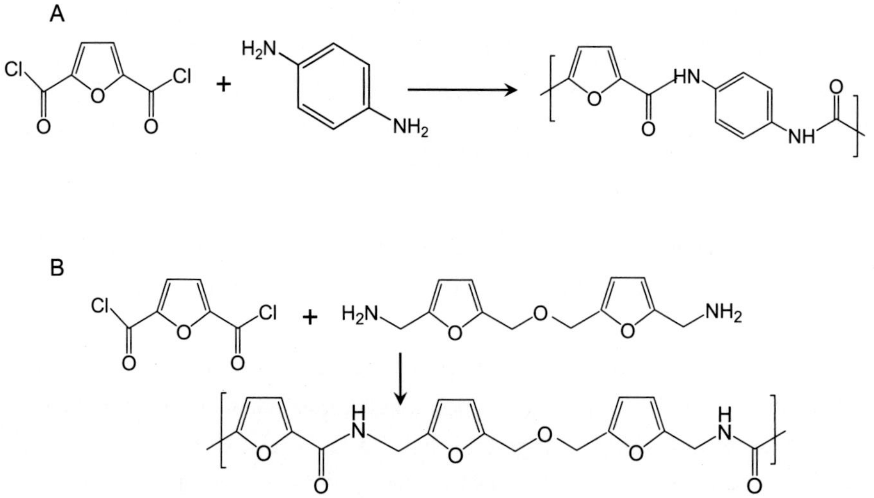

Figure 8 Possible pass way to incorporate furan derivatives into PA

from biomass has been developed. Synthesis of PA44 from succinic acid and 1,4-diamino butane which is derived from succinic acid has been reported [12].

Other potential candidates are 2-pyrone-4,6-dicarboxylic acid [13] as a diacid component. 1,5-diamino pentane (cadaverine) [14–16] and γ-aminobutyric acid [17] are also novel candidates for biobased PAs. PA56, PA510 or PA5T can be produced from 1,5-diamino pentane [18] and PA4 from γ-amino butylic acid through 2-pyrrolidone [19]. The possible ways for their production from biomass are shown in Figure 7. Lignin was oxidized under alkaline condition to vaniline and vanilic acid which were converted to 2-pyrone-4,6-dicarboxylic acid (PDC) by recombinant *Pseudomonas putida* PpY1100 in which several genes related to digestion of lignin-derived phenols were introduced. Converting yield was 95% for 24 hours.

Amino acid decarboxylase is very useful enzyme because carboxyl group is selectively removed from amino acid by this enzyme. When applying to lysine, 1,5-diamino pentane is produced. Mimitsuka *et al.* reported the production of 1,5-diamino pentane from glucose as a carbon source using recombinant *Corynebacterium glutamicum* expressing *Escherichia coli* lysine decarboxylase [14]. Yield was 2.6 g/L after 18 hours fermentation. Tateno *et al.* reported the one-step production of 1,5-diamino pentane from starch as a carbon source using a recombinant *C. glutamicum* strain coexpressing α-amylase and lysine decarboxylase [15]. Yield was 2.4 g/L after 21 hours fermentation. Kind reported the production of 1,5-diamino pentane from glucose as a carbon source using recombinant *C. glutamicum* expressing *E. coli* lysine decarboxylase [16]. Yield was 3.1 g/L after 10 hours fermentation. Conversion yield from glucose to 1,5-diamino pentane was 30%. Toray Co. Ltd., released the news about the plan of production of novel PAs based on 1,5-diamino pentane.

PA4 which is synthesized from γ-aminobutyric acid (GABA) via 2-pyrrolidone has very unique properties; high melting temperature (265°C) [19] and biodegradability [20]. The monomer, GABA, is very popular as a supplement because of its physiological function in human body such as blood pressure suppression. Many patents and papers described processes for the production of GABA using microbes such as *Lactobacillus*. *E. coli* (NBRC3806) has very high activity of glutamate decarboxylase. This strain produced 280 g of GABA from 520 g of glutamic acid. The conversion yield was 80%. The enzyme solution obtained from homogenized strains had also high activity with conversion yield of 84% after 24 hours incubation [17]. Highly thermostable glutamate decarboxylase is also useful for industrial process. Archaeal *Pyrococcus horikoshii* glutamate decarboxylase expressed by recombinant *E. coli* was purified and characterized [21].

PA4 was synthesized firstly in 1953 [22] and processes and catalysts have been improved [23]. National Institute of Advanced Industrial Science and Technology (AIST) improved molecular weight and tensile strength by using multifunctional initiators such as 1,3,5-benzenetricarboxylic acid [24]. The PA4s thus obtained have blanched structures similar to star-polymers then tensile strength increased; 50–75MPa for blanched PA4s against 15–50MPa for linear PA4s. Strain at break point is about 40–60%. Melting temperature (265 °C) can been controlled by copolymerization with ε-caprolactam to be 154 °C and strain at break point increases to be more than 500% [25].

4.2.3 Conclusion

PAs is one of widely used engineering plastics and more than 2 000 000 tons are consumed every year in the world. In order to develop biobased PAs two pathways seem to be possible: Process innovation and material innovation. In process innovation conventional monomers should be synthesized from biomass such as sugars and fats by chemical and biochemical processes. PA11 and PA1010 are typical examples. Processes for other conventional monomer production would be expected near future. In material innovation novel monomers should be synthesized from biomass and monomers should have economical advantages and same or higher performance over conventional PAs. Recently these approaches have been started such as 1,5-diamino pentane and γ-aminobutyric acid. Optimization of processes for down stream and development of new processing technology should be expected near future.

References

1. A. Inoue, *Chemical Economics* (in Japanese), **57**, 118 (2010)
2. Michigan Stated University, WO2005–123669
3. DSM, WO2006/005603, WO2006–005604
4. Purdue University, WO95–07996
5. S. K. Ritter, *Chem. Eng. News*, **82** (22), 31 (2004); J. J. Bozell and G. R. Petersen, *Green Chem.*, **12**, 539 (2010)
6. J. Lewkowski, *Polish J. Chem.*, **75**, 1943 (2001)
7. M. L. Ribeiro, U. Schuchardt, *Catal. Commun.*, **4**, 83 (2003)
8. C. Moreau, M. N. Belgacem, A. Gandini, *Top. Catal.*, **27**, 11 (2004)
9. Y. Taguchi, A. Oishi, H. Iida, *Chem. Lett.*, **37**, 50 (2008)
10. A. Gandini, *Macromolecules*, **41**, 9491 (2008)
11. A. Mitiakoudis, A. Gandini, *Macromolecules*, **24**, 830 (1991)
12. I. Bechthold, K. Bretz, S. Kabasci, R. Kopitzky, A. Springer, *Chem. Eng. Technol.*, **31**, 647 (2008)
13. Y. Otsuka, M. Nakamura, S. Ohara, Y. Katayama, K. Shigehara, E. Masai and M. Fukuda, *J. Environm. Biotechnol*, **6**, 93–103 (2006); Japan Patent Application 2007–37452A
14. T. Mimitsuka, H. Sawai, M. Hatsu, K. Yamada, *Biosci. Biotechnol. Biochem.*, **71**, 2130 (2007)
15. T. Tateno, Y. Okada, T. Tsuchidate, T. Tanaka, H. Fukuda, A. Kondo, *Appl. Microbiol. Biotechnol.*, **82**, 115 (2009)
16. S. Kind, W. Jeong, H. Schröder, C. Wittmann, *Metab. Eng.*, **12**, 341 (2010)
17. National Institute of Advanced Industrial Science and Technology, Japan Patent Application 2009–159840A
18. Toray, Japan Patent Application 2004–75962A; 2003–292613A; Mitsubishi Chemicals, Japan Patent Application 2006–348057A
19. N. Kawasaki, A. Nakayama, N. Yamano, S. Takeda, Y. Kawata, N. Yamamoto and S. Aiba, *Polymer*, **46**, 9987 (2005)
20. N. Yamano, A. Nakayama, N. Kawasaki, N. Yamamoto, S. Aiba, *J. Polym. Environ.*, **16**, 141 (2008)
21. Kim Han-Woo, K. Kashima, K. Ishikawa, N. Yamano, *Biosci. Biotechnol. Biochem.*, **73**, 224 (2009)
22. USP 2638463 (1953); USP 2739959 (1956)
23. Mitsubishi Chemical, Japan Examined patent application publication S48–42719,B; Ube Industries,

Japan Patent Application H05–295108A
24. National Institute of Advanced Industrial Science and Technology, Japan Examined patent application publication JP, 3699995, B; JP, 3453600, B
25. National Institute of Advanced Industrial Science and Technology, Japan Patent Application 2009–155608A

4. 3 Enzymatic polymerization catalyzed by hydrolases
Shiro Kobayashi

Abstract: The present chapter reviews comprehensively but briefly the method of polymer synthesis, enzymetic polymerization, in particular, with focusing the polymerization using hydrolases (hydrolysis enzymes) as catalyst, developed in these two decades. This way of writing leads to mention the synthesis mainly of polysaccharides and polyesters.

4. 3. 1 Introduction

Currently, bio-based polymers attract much attention due mainly to environmental problems; the global climate change, CO_2 emissions based on the concept of carbon-neutral, natural resouces, energy savings, and so fourth. In the polymer materials fields, these problems are within the context of "green polymer chemistry" [1–3]. Polymer production is conducted by polymerization reactions, which need catalysts (or initiators). Historically, polymerization catalysts utilized classical catalysts of acids (Brønsted acids, Lewis acids, and various cations), bases (Lewis bases and various anions), and radical generating compounds since the 1920s, the first stage of polymer chemistry. In the following second stage, the catalysts started to use the transition metals in Ziegler-Natta catalyst in the 1950s and later in the metathesis catalyst as well as the rare-earth metals. These catalysts still keep the major roles in the polymer synthesis.

Since two decades before, however, a new approach of polymer synthesis as a third stage has been developed, with employing enzymes as catalyst ("enzymatic polymerization") [2], although *in vitro* enzymatic catalysis has been extensively used in the organic synthesis area as a convenient and powerful tool [4]. Enzymatic polymerization is defined as "the *in vitro* polymerization of artificial substrate monomers catalyzed by an isolated enzyme via nonbiosynthetic (nonmetabolic) pathways". In these two decades in fact, many enzymatic polymerization reactions have been found and they brought about a variety of new polymers [2,5]. In this chapter, the new method of polymer synthesis, enzymatic polymerization, is reviewed comprehensively but briefly. In particular, enzymatic polymerization using hydrolases (hydrolysis enzymes) as catalyst, which has been the most extensively studied area, will be focused. In this way, the description automatically leads to include mainly the polysaccharide synthesis and polyester synthesis. Enzymes belong to a kind of bio-based polymers, and hence, bio-based polymers like polysaccharides and polyesters are produced with catalysis of the bio-based polymer of enzymes.

4. 3. 2 Characteristics of enzymatic reactions and basic concept of enzymatic polymerization

In living cells all biopolymers (biomacromolecules) are produced with enzymatic catalysis. Such biopolymers include nucleic acids (DNA, RNA), ptoteins, polysaccharides, polyesters, polyaromatics, natural rubber, and so fourth.

Two fundamental characteristics of enzymatic reactions must be mentioned. The first is a "key and lock" theory proposed by E. Fischer in 1894, which pointed out the relationship between an enzymatic catalysis and a natural substrate [2]. The theory implies that the enzyme catalyzes a reaction of a specific substrate that is recognized by the enzyme like a key and lock relationship as shown in the upper cycle (A) in Figure 1. The substrate is activated with forming the enzyme-substrate complex like a substrate (key) located in the enzyme (lock) with geometrical adaptation to lead to the product via the bond formation.

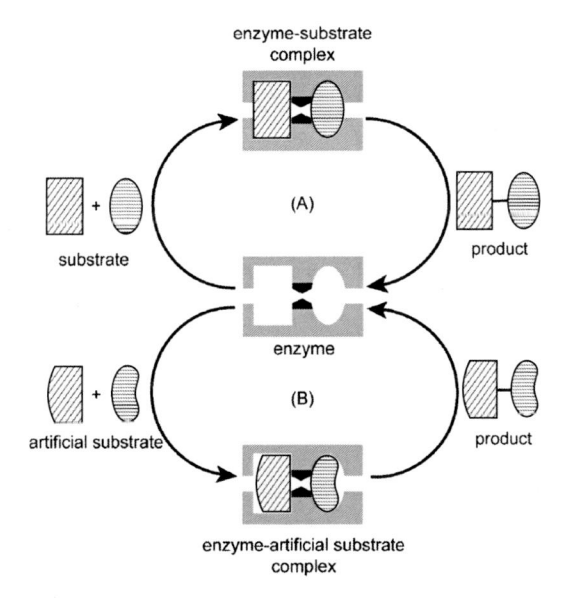

Figure 1 Enzyme-substrate relationship for enzymatic reactions
(A) An *in vivo* reaction obeying the "key and lock" theory. (B) An *in vitro* reaction involving an enzyme-artificial substrate complex leading to a product. The black part in the enzyme indicates the active site.
(Reprinted with permission from ref. 2d. Copyright 2009 American Chemical Society.)

The second is also concerning the enzymatic reaction mechanism. L. Pauling suggested the reason why an enzymatic reaction progresses under mild reaction conditions. The activation energy is much lowered by stabilizing the *transition-state* involving an enzyme-substrate complex with comparison to the no-enzyme reaction (Figure 2) [2]. The rate acceleration is normally 10^6–10^{12} folds; however, a specific case reached even 10^{20} folds! [6] The mechanism of *in vivo* enzymatic reaction shown in Scheme 1 is generally accepted, which corresponds to the reaction coordinate in Figure 2.

The *in vitro* enzymatic reaction lies on the following rationale. The "key and lock" relationship observed for *in vivo* reactions is not absolutely strict for enzymatic reactions. Enzyme is often dynamic and able to interact with not only a natural substrate but also an unnatural substrate. In the *in vitro* enzymatic polymerization, a monomer is an unnatural (artificial) substrate for the

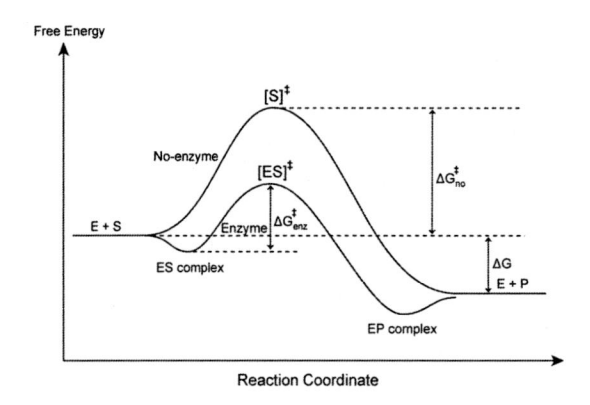

Figure 2 Energy diagram for a chemical reaction
An enzyme-catalyzed reaction proceeds much faster than a no-enzyme reaction, by lowering the activation energy with stabilizing the transition-state of the reaction. (Reprinted with permission from ref. 2d. Copyright 2009 American Chemical Society.)

$$E + S \rightleftharpoons ES \rightleftharpoons [ES]^{\ddagger} \rightleftharpoons EP \rightleftharpoons E + P$$

E: enzyme, S: substrate, P: product

Scheme 1

catalyst enzyme. Yet, the substrate is to be recognized and to form an enzyme-artificial substrate complex so that the desired reaction may take place. It has been proposed, therefore, that the monomer is to be designed according to a new concept of a "transition-state analogue substrate" (TSAS), the structure of which should be close to that of the transition-state of the *in vivo* enzymatic reaction [2,4,7,8]. This is due to the fact that the enzyme stabilizes the transition-state via complex formation with the substrate [2]. An appropriately designed artificial monomer thus forms readily an enzyme-substrate complex and the reaction is induced to give the product with liberating the enzyme again as shown in the lower cycle (B) in Figure 1. It is stressed that a structurally resembling transition-state must be involved commonly in both cycles (A) and (B).

In vivo enzymatic reactions involve following characteristics: (i) high catalytic activity (high turnover number), (ii) reactions under mild conditions with respect to temperature, pressure, solvent, pH of medium, *etc*., bringing about energetic efficiency, and (iii) high reaction selectivity of regio-, enantio-, chemo-, and stereo-regulations, giving rise to perfectly structure-controlled products. If these *in vivo* characteristics could be realized for *in vitro* enzymatic polymer synthesis, we may expect the following outcomes: (a) perfect control of polymer structures, (b) creation of polymers with a new structure, (c) a clean, selective process without forming by-products, (d) a low loading process with saving energy, and (e) biodegradable properties of product polymers in many cases. These are indicative of "green" nature of the *in vitro* enzymatic reactions for developing new polymeric materials. Actually, many of the expectations have been realized as seen below.

All enzymes are classified into six main groups according to the Enzyme Commission (Table

1). Today, many thousands of enzymes are commercially available and some of them are mutated for industrial applications. Generally, oxidoreductases, hydrolases, and isomerases are relatively stable and some of isolated enzymes among them are conveniently used as catalyst practically like in chemical-, food-, and pharmaceutical industries. In contrast, lyases and ligases are in less amount in living cells and less stable for isolation or separation from living organisms. So far, the first three groups of enzymes have been employed as catalyst for enzymatic polymerization to produce various polymers as shown in Table 1.

Table 1   Classification of enzymes, their examples, and typical polymers synthesized by *in vitro* enzymatic catalysis

enzymes	example enzymes	synthesized polymers
1. Oxidoreductases	peroxidase, laccase, tyrosinase, glucose oxidase	polyphenols, polyanilines, vinyl polymers
2. Transferases	phosphorylase, glycosyltransferase, acyltransferase	polysaccharides, cyclic oligosaccharides, polyesters
3. Hydrolases	glycosidase (cellulase, amylase, chitinase, hyaluronidase), lipase, peptidase, protease	polysaccharides, polyesters, polycarbonates, polyamides, polyphosphates, polythioesters
4. Lyases	decarboxylase, aldolase, dehydratase	
5. Isomerases	racemase, epimerase, isomerase	
6. Ligases	ligase, synthase, acyl CoA synthetase	

4. 3. 3   Synthesis of polysaccharides

Polysaccharides include one of three major classes of natural biomacromolecules together with nucleic acids and proteins. The synthesis of two other major biomacromolecules has been facilitated for many years by utilizing an automated solid-phase synthesizer based on the Merrifield method and/or a genetic engineering procedure. In contrast, polysaccharides have very complicated structures, having many stereo- and regio-isomers, and hence, it had been so hard to develop a versatile synthesis method of polysaccharides. In these two decades, however, a new method of synthesizing structurally well-defined polysaccharides has been developed by using hydrolases (hydrolysis enzymes) as catalyst ("enzymatic polymerization") [2,5,8].

The polysaccharide synthesis needs a highly selective glycosylation reaction between a donor molecule and an acceptor molecule to form a glycosidic linkage as given in a simple example of $\beta(1\rightarrow4)$ linkage formation (Scheme 2). This type of glycosylation reaction must occur repeatedly many times for producing a higher molecular weight polysaccharide.

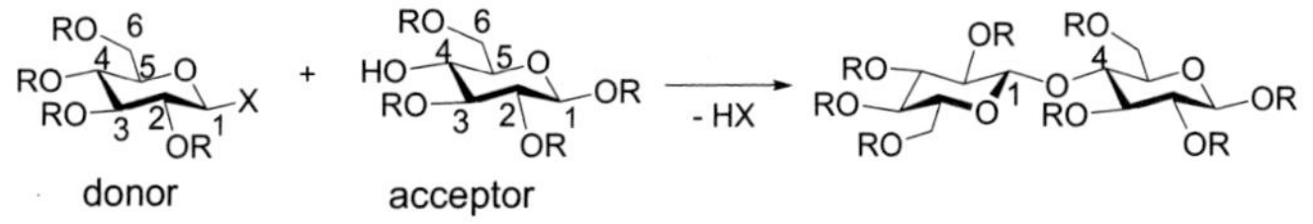

Scheme 2

Enzymatic reactions are very specific to proceed in regio-selective and stereo-controlled manner. In the biosynthetic pathway of natural polysaccharides, formation of a glycosidic linkage is mainly catalyzed by glycosyltransferases employing the corresponding sugar nucleotides as substrate. For example, natural cellulose is synthesized *in vivo* from uridine-5′-diphospho (UDP)-glucose via cellulose synthase-catalyzed reaction. However, glycosyltransferases are generally transmembrane-type proteins, present in nature in very small amount, and unstable for isolation and purification with difficulty in handling, and hence, it is not easy to apply a glycosyltransferase for an *in vitro* reaction.

As can be speculated from Scheme 1, all the enzymatic reactions are reversible in principle. Accordingly, in our challenge to synthesize polysaccharides, a hydrolysis enzyme was employed for the glycosidic-linkage formation [2,7]. Glycosidase is a hydrolysis enzyme, which catalyzes cleavage reaction of glycosidic linkages *in vivo*.

Natural polysaccharides and their derivatives by polycondensation

Cellulose. Cellulose has a linear structure of a $\beta(1\rightarrow4)$ linked D-glucose (Glc) repeating unit, and is found as by far the most abundant organic compound on the earth. It is one of the three major structural components of the primary cell walls of green plants, along with hemi-cellulose and lignin.

In spite of the importance of cellulose, an *in vitro* chemical synthesis of cellulose had not been achieved since the first challenge for this in 1941.

In 1991, the first *in vitro* synthesis of cellulose was reported for the polymerization of β-cellobiosyl fluoride (β-CF) monomer catalyzed by cellulase, a hydrolysis enzyme, from *Trichoderma viride* [7]. Cellulase is an enzyme, which catalyzes hydrolysis reaction of $\beta(1\rightarrow4)$ glycosidic linkage between two Glc units *in vivo*. In an acetonitrile/acetate buffer (pH 5.0) (5:1, v/v) mixed solvent, cellulase catalyzed the bond-formation reaction, and produced "synthetic cellulose" having a perfectly controlled $\beta(1\rightarrow4)$ glycosidic structure with degree of polymerization (DP) value around 22 (Scheme 3). During the reaction HF was liberated and hence the reaction is of polycondensation.

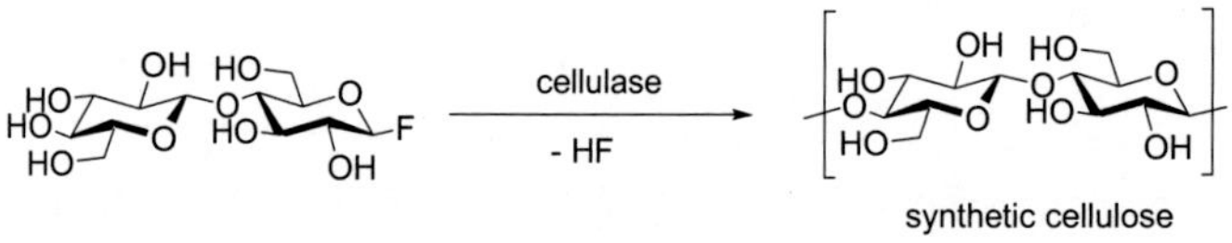

Scheme 3

Figure 3 illustrates a proposed mechanism of the cellulose synthesis [2,5d,5f,7,8]. In the hydrolysis (A), one acid residue protonates glycosidic oxygen atom, and the other pushes the anomeric (C-1) carbon in the general acid-base mode, enabling the cleavage of the glycosidic linkage (stage a). Then, a highly reactive intermediate (or transition-state) of a glycosyl-carboxyl structure with α-configuration is formed (stage b). To the anomeric center of the intermediate, water molecule attacks from the β-side, and hydrolysis is completed (stage c). Therefore, cellulase is a retaining enzyme with "double displacement mechanism", which involves twice inversions of anomeric stereochemistry. In the polymerization (B), immediately after the recognition of β-CF monomer at the donor site of cellulase catalyst, fluoride anion is readily eliminated as HF molecule via general acid-base mechanism (stage a'). The monomer forms a glycosyl-carboxyl intermediate with α-configuration, whose structure is regarded as similar with that of stage b of the hydrolysis (stage b'). C-4 Hydroxyl group of the disaccharide monomer or the growing chain end located at the acceptor site attacks the anomeric carbon from the β-side, and new β(1→4) glycosidic linkage is formed (stage c'). Thus, β-CF is considered as a "transition-state analogue substrate" (TSAS) monomer because of a transition-state (or intermediate) structure involved in common in both reactions from stage a (a') to stage b (b'). During the polymerization, this glycosydic linkage formation reaction is repeated, and hence, the monomer acted as a glycosyl donor as well as a glycosyl acceptor.

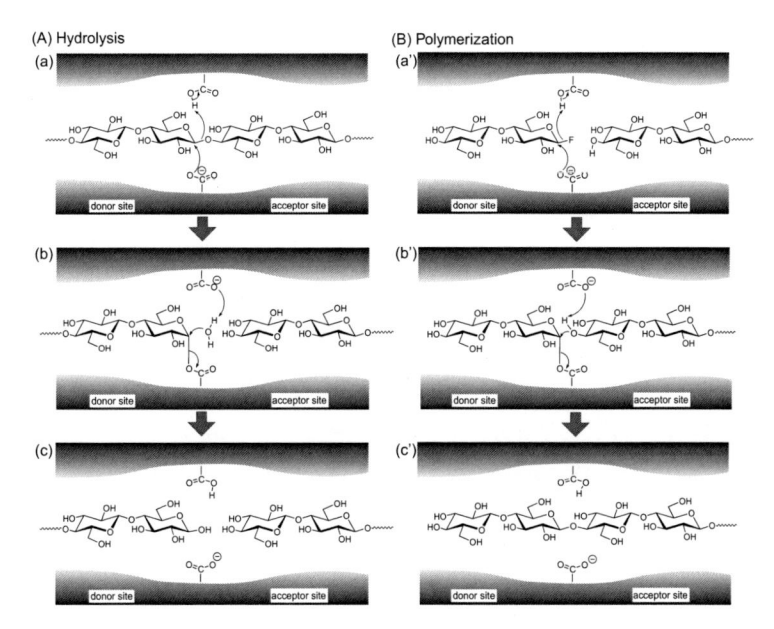

Figure 3 Postulated reaction mechanisms of cellulase catalysis for hydrolysis of cellulose (A) and for polycondensation of β-CF to synthetic cellulose (B). (Reprinted with permission from ref. 8. Copyright 2007 The Japan Academy.)

β-CF is excellent as a TSAS monomer, and the reason is considered as the following four factors; (1) a disaccharide structure of cellobiose is the smallest unit of the cellulose repeating unit structure to be recognized by cellulase, (2) the size of fluorine atom (covalent radius, 0.64 Å) is close to that of oxygen (0.66 Å), giving a desired opportunity for β-CF to be recognized by cellulase, (3) fluoride anion is the best leaving group, and (4) only the glycosyl fluoride is stable among the unprotected glycosyl halides.

Another approach of cellulose synthesis was recently achieved, extending the above method in principle which utilized a disaccharide monomer of cellobiose and a cellulase/surfactant (CS) complex as catalyst [9]. The CS complex was consisted from the mixture of cellulase and a specific nonionic surfactant of dioleyl-N-D-glucona-L-glutamate ($2C_{18}\Delta^9$GE). In the nonaqueous medium of dimethylacetamide (DMAc)/LiCl, a cellulose-solubilizing solvent, polymerization took place at 37°C to give white powders as synthetic cellulose. The DP value was high (over 100) and the product yield was low (up to 5%). The reaction is of dehydration polycondensation and controlled regioselectivity and stereochemistry, involving an extensive transglycosylation of the product with a statistical even and odd numbered DP.

The cellulose synthesis was achieved by using a cellulose-binding domain (CBD)-deleted mutant cellulase as catalyst. A problem of the cellulase-catalyzed synthesis is the hydrolysis of the product cellulose, which was considered to be suppressed via this method. The mutant enzyme of cellulase (endoglucanase II, EGII) lacking the CBD was expressed in yeast (Figure 4). The enzyme, $EGII_{(core)}$, catalyzed the polymerization of β-cellobiosyl fluoride (β-CF) to form synthetic cellulose with $\beta(1 \rightarrow 4)$ glycosidic linkage. As expected the resulting crystalline product was hardly hydrolyzed by the mutant enzyme [10a]. This result indicated that CBD is not important for inducing the polymerization. A new mutant enzyme, $EGII_{(core)2}$, having sequential two active sites of EGII was prepared. $EGII_{(core)2}$ enzyme showed higher polymerization and hydrolysis activities than $EGII_{(core)}$. This was considered mainly due to the suitably stabilized conformation with the sequential arrangement [10b].

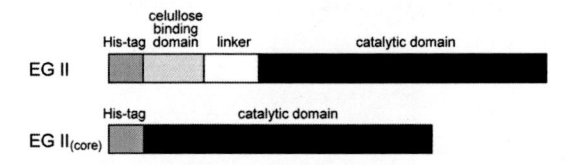

Figure 4 Schematic representation of EG II expressed by *Saccaromyces serevisiae* and the mutant EG II$_{(core)}$.
(Reprinted with permission from ref. 8. Copyright 2007 The Japan Academy.)

Chemical synthesis of cellulose based on a conventional organic technique is to be mentioned as reference, which was archived in 1996 via cationic ring-opening polymerization of 3,6-di-O-benzyl-α-D-glucose 1,2,4-orthopivalate [11]. After removing the protecting groups from the polymerization product, it was claimed to be synthetic cellulose with a similar DP value with that obtained by the enzymatic method.

As to the high-order self-assembling molecular structure, there are typically two types of allomorphs in cellulose. One is thermodynamically metastable cellulose I, in which cellulose chains are aligned in parallel. The other is thermodynamically stable antiparallel cellulose II. Surprisingly, naturally occurring cellulose forms less stable cellulose I crystalline structure. *In vitro*, crystalline structures of cellulose synthesized via cellulase-catalyzed polymerization were cellulose II with crude enzyme and cellulose I with purified enzyme. Metastable cellulose I could be formed due to the kinetically controlled process [12]. This was the first instance of cellulose I formation via a non-biosynthetic pathway. Such control in high-order molecular assembly during polymerization was not observed before, therefore, a new concept of "choroselective polymerization" was developed [13].

A self-assembling process of synthetic cellulose during cellulase-catalyzed polymerization was investigated in detail at real time and in-situ by a combined small-angle scattering (SAS) method, [combining small-angle neutron scattering (SANS), small-angle X-ray scattering (SAXS), ultra-SANS and ultra SAXS methods], together with wide-angle X-ray scattering (WAXS) and field-emission scanning electron microscopy (FE-SEM) [14]. It was disclosed that cellulase aggregates into associations with characteristic lengths larger than 200 nm in aqueous reaction medium. Further, cellulose molecules created at each active site of enzymes associate themselves around the enzyme associations into cellulose aggregates having surface fractal dimensions D_s, increasing from 2 (smooth surface) to 2.3 (rough surface with fractal structure) with the reaction time progress, extending over a surprisingly wide length scale ranging from ~30 nm to ~30 μm with three orders of magnitude. This unique self-assembly is caused by an extremely large number of cellulose molecules repeatedly created at the active center of the enzyme.

Cellulose Derivatives. Syntheses of C-6 methylated cellulose derivatives were achieved via cellulase-catalyzed polymerization of 6-O-methyl- and 6'-O-methyl-cellobiosy fluorides (Scheme 4) [15]. 6-O-Methyl-β-cellobiosy fluoride was recognized by cellulase as a TSAS, giving rise to the corresponding alternatingly C-6 methylated cellulose derivative. The number-average molecular weight (M_n) of the product was 3.9×10^3, corresponding to n = 7. On the other hand, only dimer (tetrasaccharide) was obtained from 6'-O-methyl-β-cellobiosy fluoride.

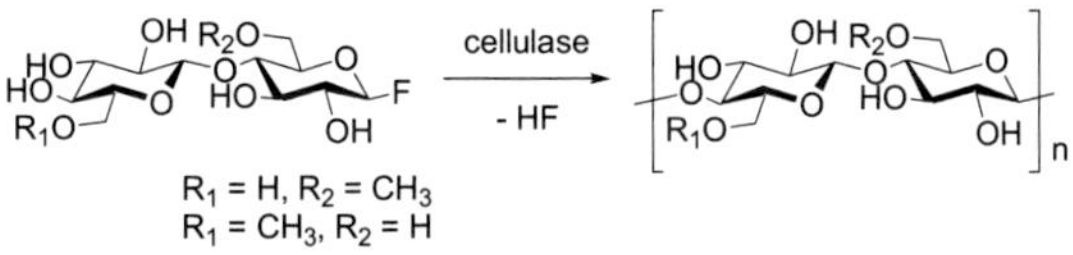

Scheme 4

Amylose. Amylose is a polymer of glucose linked through $\alpha(1\rightarrow4)$ glycoside bonds. α-Amylase-catalyzed polycondensation of α-D-maltosyl fluoride in a methanol-phosphate buffer (pH 7.0), 2:1 (v/v) mixed solution gave amylose oligomers up to heptasaccharide (Scheme 5) [16]. Oligomer formation may be due to a steric hindrance caused from the helical structure of amylose as barrier for producing longer molecules.

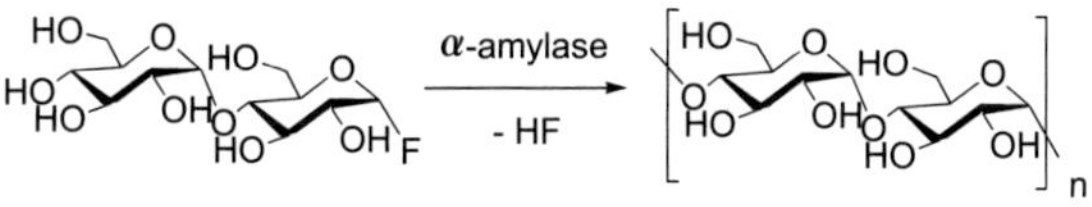

Scheme 5

Phosphorylase is an exo-type enzyme, which catalyzes *in vivo* phosphorolysis at the non-reducing end of the glycosidic linkage. Phosphorylase is not a hydrolysis enzyme but a transferase enzyme. Its catalysis is noticeable and hence mentioned here. Amylose was synthesized *in vitro* by the phosphorylase-catalyzed polymerization of α-D-glucose 1-phosphate (Glc-1-P) monomer, which was initiated from a primer of maltoheptaose [17]. When the polycondensation was carried out in the presence of polyTHF as a hydrophobic polymer, the polymerization proceeded with the formation of the amylose-polyTHF inclusion complex (Scheme 6) [18]. Formation of amylose-polymer inclusion complexes was conducted with various axle hydrophobic polymers. This unique polymerization to produce amylose-polymer inclusion complexes was termed as "vine-twining polymerization" [19].

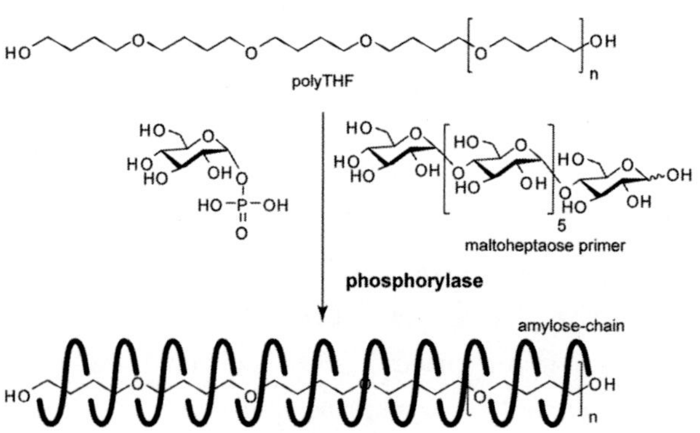

Scheme 6

Xylan. Xylan is one of the most important components of hemicelulose in plant cell walls, being a polysaccharide of xylose having a β(1→4) glycosidic linkage. Based on the TSAS concept, β-xylobiosyl fluoride was designed as monomer and subjected to polycondensation catalyzed by cellulase (Scheme 7). The product was a synthetic xylan having β(1→4) glycosidic linkage, with an average molecular weight of 6.7×10^3, a DP value of ~23 [20].

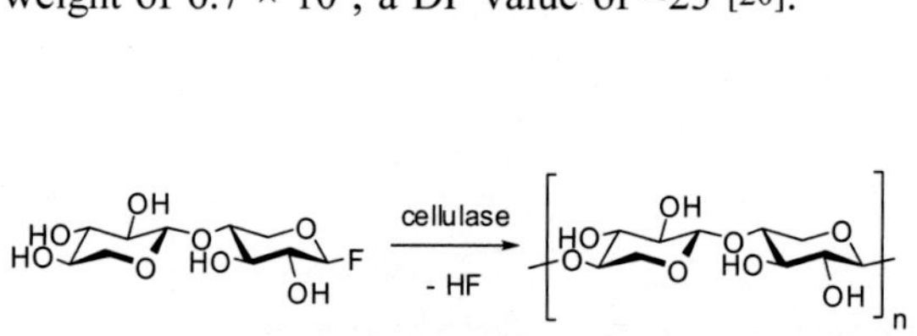

Scheme 7

Natural polysaccharides and their derivatives by ring-opening polyaddition

Chitin. Chitin is a β(1→4) linked *N*-acety-D-glucosamine (GlcNAc) polysaccharide, which is synthesized *in vivo* by the catalysis of chitin synthase with UDP-GlcNAc as a substrate. Chitosan is an *N*-deacetylated product of chitin. They are one of the most abundant and widely found polysaccharides in the animal field as a structural material, and show excellent characteristics of biodegradability, biocompatibility, and in particular low immunogenicity.

In 1995 the first *in vitro* synthesis of chitin was accomplished via ring-opening polyaddition of a chitobiose oxazoline derivative monomer (a di-saccharide) catalyzed by chitinase, and published in 1996 (Scheme 8) [21a–c]. The reaction is of ring-opening polyaddition mode, and was promoted under weak alkaline conditions (pH 9.0–11.0), where the hydrolysis of the product chitin was much suppressed. The DP value of "synthetic chitin" was evaluated as 10–20 depending on the reaction conditions [8].

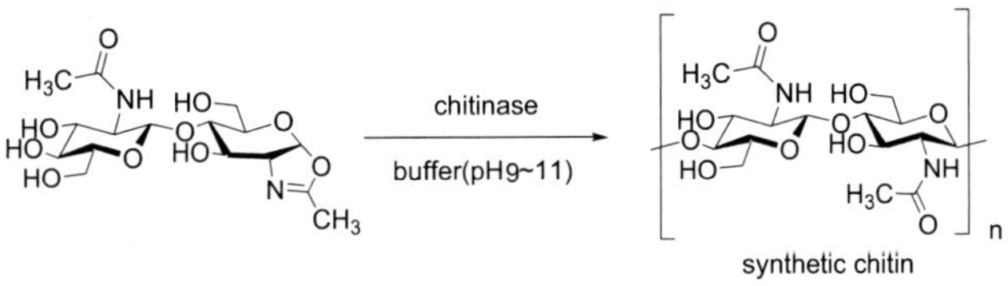

Scheme 8

As to the mechanism of the chitinase catalysis, two carboxylic acid residues are involved for the catalysis (Figure 5). In the hydrolysis of chitin, oxygen atom of glycosidic linkage is protonated by one of the carboxylic residues immediately after the recognition of chitin (stage a). Carbonyl oxygen atom of C-2 acetoamide group is attacked by the neighboring anomeric carbon from the α-side to form an oxazolinium stabilized by another carboxylic residue, and the C-1-O bond cleavage is completed (stage b). Therefore, the reaction is of "substrate assisted mechanism" [8]. Ring-opening of this oxazolinium ring occurs by nuclephilic attack of water molecule from β-side, giving rise to the hydrolysate in a retaining manner (stage c). It is considered that the oxazolinium ion formation is a transition-state (or intermediate) in the hydrolysis. In the polymerization, the monomer is first recognized and at the same time the nitrogen atom is protonated by the carboxylic acid residue at the donor site of chitinase, forming the corresponding oxazolinium ion (stage b'). This is because the monomer has already an oxazoline structure. Instead of the water molecule in the hydrolysis, C-4 hydroxyl group of sugar unit located at the acceptor site attacks nucleophilically the anomeric carbon from the β-side to open the ring, providing a new glycosidic linkage with β-configuration (step c'). Thus, the monomer acted as a glycosyl donor as well as a glycosyl acceptor, and this glycosidic linkage formation reaction is repeated during the polymerization. The most important point is the structure resemblance of the transition-state (or intermediate) involved in both reactions at stage b and stage b'. From these considerations a new concept of TSAS was proposed [21a], and then a mechanism involving an oxazolinium ion intermediate in the chitin hydrolysis by family 18 chitinase was reported [21b]. During the polymerization, 2-methyl-oxazoline ring of the monomer played a role of forming *N*-acetyl group by opening the ring; the ring can be taken as a masked *N*-acetyl group.

The polymerization was extended to a step-wise elongation of GlcNAc unit, which was

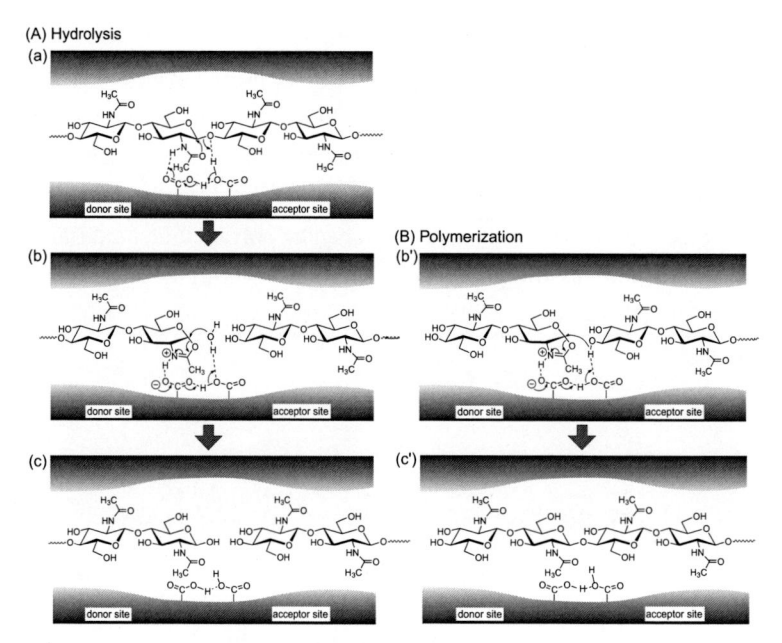

figure 5 Proposed mechanisms of chitinase catalysis for hydrolysis of chitin (left) and for ring-opening polyaddition of the chitobiose oxazoline monomer to synthetic chitin (right).
(Reprinted with permission from ref 8. Copyright 2007 The Japan Academy.)

performed via combined use of chitinase and β-galactosidase. Chitinase recognized 4-OH group of GlcNAc as an acceptor. Therefore, by the catalysis of the enzyme, no self-polyaddional products were produced from N-acetyllactosamine (LacNAc) oxazoline derivative, a disaccharide of galactose (Gal)β(1→4)GlcNAc. Chitinase catalyzed the glycosylation of LacNAc oxazoline derivative to the 4-OH group of GlcNAc acceptor. After the glycosylation, Gal unit located at non-reducing end of the product was removed via β-galactosidase-catalyzed Galβ(1→4)GlcNAc glycosidic bond cleavage. Repetitions of these sequential manipulations enabled to synthesize the chain-length controlled chitooligosaccharides [22].

Chitin Derivatives. As new TSAS monomers 3-*O*-methyl, and 3'-*O*-methyl-chitobiose oxazoline derivatives were designed and both monomers produced only chitooligosaccharides in rather low yields [23]. 6-*O*- or 6'-*O*-Carboxylmethylated (CM) chitobiose oxazoline derivative monomers were polymerized by chitinase, indicating that both monomers were recognized by donor site of chitinase. However, resulting product was dimer (tetrasaccharide) from the reaction of 6'-*O*-CM chitobiose oxazoline derivative [24].

Fluorine-substituted chitobiose oxazoline derivatives having C-6 fluorine were prepared because of the similar size of the fluorine atom to oxygen atom, and hence similar steric factor between OH and F as subtituent at C-6. Under the weak alkaline conditions, reactions of these monomers were induced by chitinase, giving rising to the white precipitates of structurally well-defined fluorinated-chitins with the average molecular weights of F-chitins of 1,400–1,700 [25]. A TSAS monomer bearing a bulky *N*-sulfonate group at the C-2' position was designed. Chitinase-catalyzed polymerization proceeded homogenously, due to a good solubility of the resulting polysaccharide. By selecting appropriate chitinase enzymes, molecular weight was controlled to some extent. For example, chitinase from *Serratia marcescens* provided a polysaccharide of M_n 4,180 [26].

With copolymerization of *N*-acetylchitobiose oxazoline monomer with *N, N'*-diacetylchitobiose oxazoline monomer, tailor-made synthesis of a chitin derivative with controlled deacetylated extent ranging from 0% to 50% was conducted [27].

Glycosaminoglycans. Glycosaminoglycans (GAGs) are polysaccharides, usually linking to various proteins to form proteoglycans. Proteoglycans, collagens, and fibronectins fill the interstitial space between living cells, named extracellular matrices (ECMs), and act as a compression buffer against the stress placed in the ECMs. GAGs play important roles in living organisms, and are frequently used as therapeutic materials and food supplements. GAGs are hetero-polysaccharides consisted from a hexosamine and an uronic acid (Figure 6) [8].

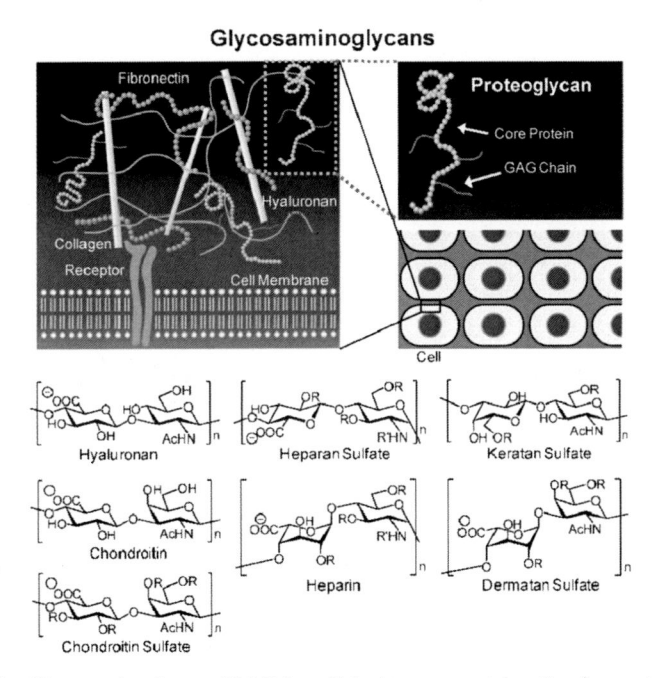

Figure 6 Glycosaminoglycans (GAGs) are linked to core proteins, forming proteoglycans. Chemical structures of typical seven GAGs are given.
(Reprinted with permission from ref. 8. Copyright 2007 The Japan Academy.)

Hyaluronan. Hyaluronan (hyaluronic acid, HA) is a non-sulfated GAG and one of the major component of ECMs. HA plays important functions *in vivo*. HA has a disaccharide repeated structure of D-glucronic acid (GlcA)-β(1→3)-*N*-acetyl-D-glucosamine (GlcNAc) connecting through β(1→4) glycosidic linkage. Hyaluronidase (HAase) is an endo-type glycoside hydrolase, which hydrolyzes β(1→4) glycosidic linkage between GlcNAc and GlcA. A novel GlcAβ(1→3)GlcNAc oxazoline derivative was designed as a TSAS monomer and its HAase-catalyzed polymerization induced the polymerization of the monomer, giving rise to HA with M_n value 1.74×10^4 in 52% yields. Regio- and stereochemistry of the resulting HA was perfectly controlled ("synthetic hyaluronan") (Scheme 9) [28]. The hydrolysis mechanism of HAase is considered to involve a double displacement mechanism, which is similar to that of chitinase. Synthetic HA seems to be an example polymer having one of the most complicated structures ever prepared *in vitro*.

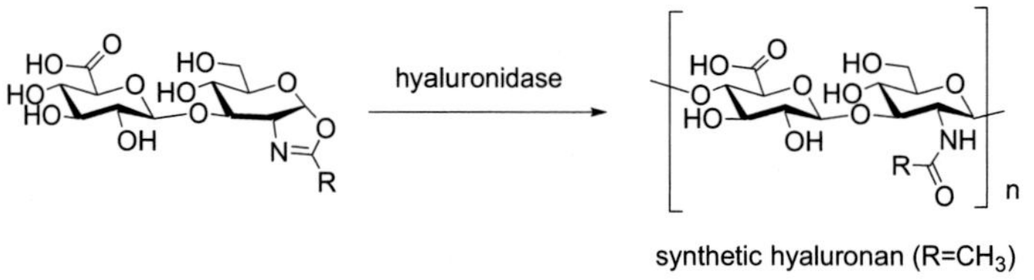

Scheme 9

Some HA derivatives having R group like ethyl, *n*-propyl and vinyl groups have been derived. The HA derivative having *N*-acryloyl group has a reactive vinyl group, and hence, functions as maromonomer, leading eventually to telechelics and graft copolymers [29].

Chondroitin. Chondroitin (Ch) is a sulfated GAG composed of alternatingly aligned *N*-acetyl-D-galactosamine (GalNAc) and D-glucronic acid (GlcA). Chondroitin sulfate (ChS) in nature is known to contain four kinds according to the site of sulfation. Ch sulfated at C-4 of GalNAc, C-6 of GalNAc, C-2 of GlcA and C-6 of GalNAc, and C-4 and C-6 of GalNAc are named as ChS-A, ChS-C, ChS-D, and ChS-E, respectively. Many papers have been published, describing the functions of ChS at a molecular level and the various sulfation patern confers specific biological activities.

A new oxazoline derivative (R = 2-methyl) as well as 2-ethyl, 2-*n*-propyl, 2-isopropyl, 2-phenyl, and 2-vinyl oxazoline derivatives were designed as new TSAS monomers for hyaluronidase (HAase) (Scheme 10) [30]. The polymerization of GlcAβ(1→3)GalNAc oxazoline monomer (R = CH$_3$) catalyzed by HAase produced the corresponding non-sulfated Ch. The M_n value of "synthetic chondroitin" reached 5.0×10^3, corresponding to that of naturally occurring Ch. Unnatural *N*-propionyl and *N*-acryloyl derivatives of Ch were obtained, with M_n of 2,700 and with M_n of 3,400, respectively.

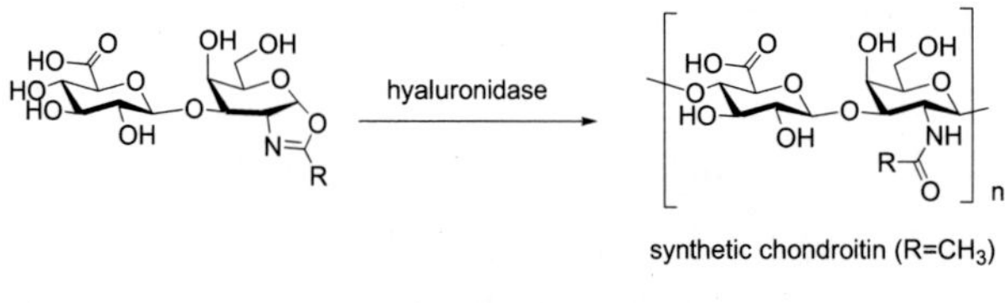

Scheme 10

Structurally well-defined ChS was prepared using HAase catalyst. Three oxazoline monomers sulfated at C-4 of GalNAc, C-6 of GalNAc, and C-4, 6 of GalNAc were synthesized [31]. Among them. a TSAS monomer sulfated at C-4 gave rise to the ChS in good yields. The resulting ChS has sulfonate group exclusvely at C-4 of GalNAc unit, a pure ChS-A. The M_n value ranged from 4.0×10^3 to 1.8×10^4.

Keratan Sulfate. Keratan sulfate (KS) is one of the class of GAGs, having a repeated disaccharide structure of $\beta(1\rightarrow 3)$ linked Gal$\beta(1\rightarrow 4)$GlcNAc (LacNAc). Gal$\beta(1\rightarrow 4)$GlcNAc(6S) and Gal(6S)$\beta(1\rightarrow 4)$GlcNAc(6S) oxazoline monomers were designed as new TSAS monomers for keratanase II. These two sulfated monomers were polymerized by keratanase II, producing the corresponding KS oligosaccharides [32].

Hybrid type unnatural polysaccharides

With extending the wide spectrum in substrate recognition of glycoside hydrolases, syntheses of unnatural polysaccharides composed from different two polysaccharide components were achieved [33]. Such polysaccharides (hybrid polysaccharides) are very difficult to synthesize via conventional chemical synthesis.

Cellulose-Xylan Hybrid Polysaccharide. The first example of unnatural hybrid polysaccharide is a cellulose-xylan hybrid polysaccharide [34]. There are two possible candidate monomers are Glc$\beta(1\rightarrow 4)$Xyl-β-fluoride and Xyl$\beta(1\rightarrow 4)$Glc-β-fluoride. Both monomers were polymerized by xylanase from *Trichoderma viride*, giving rise to the corresponding polysaccharide in a 5:1 (v/v) acetonitrile-buffer solution.

Cellulose-Chitin Hybrid Polysaccharides. Two kinds of monomers of Glc$\beta(1\rightarrow 4)$GlcNAc oxazoline and GlcNAc$\beta(1\rightarrow 4)$Glc-β-fluoride were possible to design based on the TSAS concept for chitinase and cellulase catalyses, respectively (Scheme 11(a) and 11(b)) [35]. Both enzymes catalyzed the polymerization, (a) via ring-opening polyaddition and (b) via polycondensation, affording the cellulose-chitin hybrid polysaccharide as white precipitates. The M_n values of the products reached 4.0×10^3 and 2.8×10^3, respectively. Despite a high crystalline structure of both cellulose and chitin, the cellulose-chitin hybrid showed no crystalline structure.

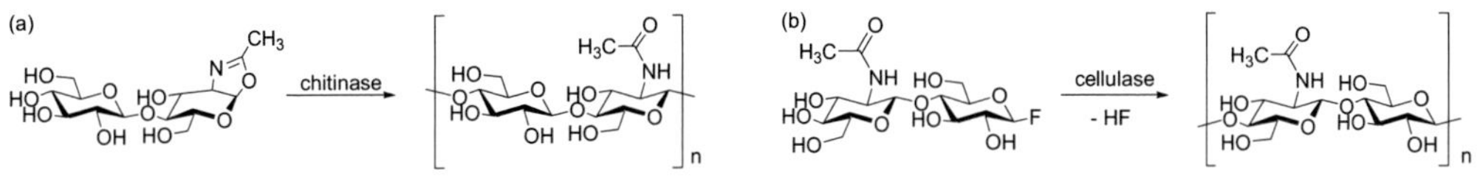

Scheme 11

Chitin-Xylan Hybrid Polysaccharide. Synthesis of a chitin-xylan hybrid polysaccharide was performed using a Xyl$\beta(1\rightarrow 4)$GlcNAc oxazoline TSAS monomer with chitinase catalyst [36]. The resulting polysaccharide showed good solubility in water and no precipitate was observed during the polymerization. The detectable highest peak was the sugar unit of 60, corresponding to a mass weight larger than 1.0×10^4.

Chitin-Chitosan Hybrid Polysaccharide. A new *N*-acetylchitobiose oxazoline derivative, whose non-reducing end of the *N*-acetyl group was deacetylated, was polymerized by chitinase catalyst to produce the corresponding polysaccharide having alternatingly aligned *N*-deacetylated group. Chitosan is an *N*-deacetylated product of chitin, which has a $\beta(1\rightarrow 4)$ linked D-glucosamine (GlcN) structure. Therefore, the resulting polysaccharide is a "chitin-chitosan hybrid polysaccharide" (Scheme 12). During the polymerization, the reaction mixture was homogeneous. A chitin-chitosan hybrid polysaccharide with $M_n = 2,020$ was obtained in 75% yields catalyzed by *Serratia marcescens* [37].

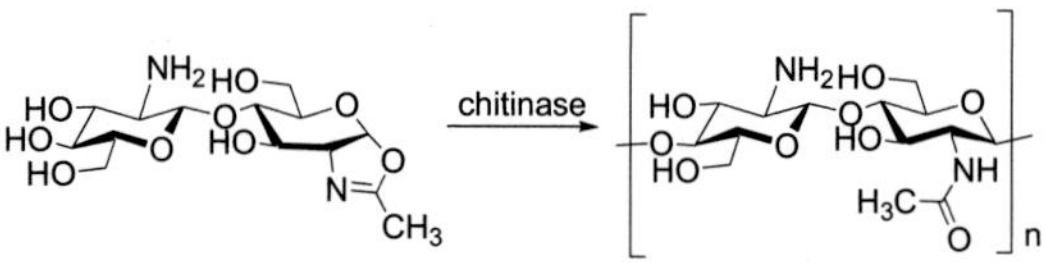

Scheme 12

Glycosaminoglycan Hybrid Polysaccharides. The synthesis of GAGs hybrids via HAase-catalyzed polymerization was conducted. Via the copolymerization of GlcAβ(1→3)GlcNAc oxazoline monomer with GlcAβ(1→3)GalNAc oxazoline monomer, a copolymer with M_n 7.4 × 10^3 was obtained in 50% yields. In a similar way, a copolymer with M_n 1.4 × 10^4 was produced in 60% yields from the copolymerization with the oxazoline monomer sulfated at C-4 of GalNAc. Copolymer composition was controlled by changing the monomer feed ratio [38].

4. 3. 4 Synthesis of polyesters

Polyesters are very important materials widely used actually like poly(ethylene terephtalate) (PET) [an aromatic polyester], poly(butylene succinate) (PBS), poly(ε-caprolactone) (PCL or poly(ε-CL)), and poly(lactic acid) (PLA) [aliphatic polyesters] (Scheme 13). Industrially, the former two are produced via polycondensation, and the latter two via ring-opening polymerization.

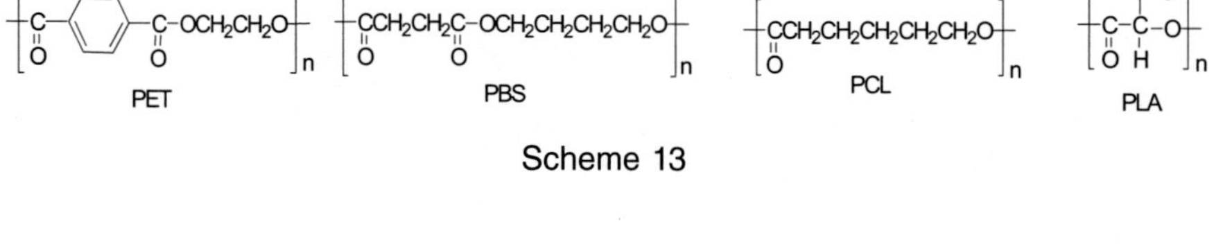

Scheme 13

Lipases (triacylglycerol acylhydrolase) catalyze the hydrolysis of fatty acid glycerol esters *in vivo* with bond-cleaving. As a hydrolase enzyme induced the polymerization to yield polysaccharides described in the preceding section, lipase catalyzes polymerization reactions to give polyesters *in vitro* with bond-forming. In fact, a variety of polyester synthesis reactions have been developed in these two decades [2,5]. The reactions proceed via two major polymerization modes; (1) polycondensation between a carboxyl group and an alcohol group, being divided into sub-modes (a) and (b), and (2) ring-opening polymerization (Scheme 14).

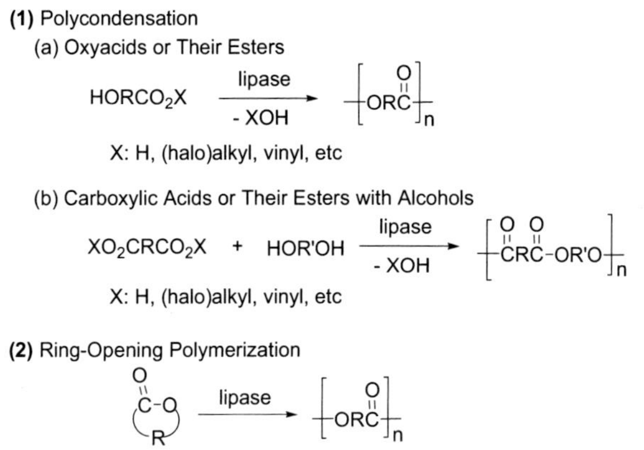

(1) Polycondensation
(a) Oxyacids or Their Esters

$$\text{HORCO}_2\text{X} \xrightarrow[-\text{XOH}]{\text{lipase}} \text{-[ORC(=O)]}_n\text{-}$$

X: H, (halo)alkyl, vinyl, etc

(b) Carboxylic Acids or Their Esters with Alcohols

$$\text{XO}_2\text{CRCO}_2\text{X} + \text{HOR'OH} \xrightarrow[-\text{XOH}]{\text{lipase}} \text{-[C(=O)RC(=O)-OR'O]}_n\text{-}$$

X: H, (halo)alkyl, vinyl, etc

(2) Ring-Opening Polymerization

$$\underset{R}{\overset{O}{\underset{\|}{C}}\text{-O}} \xrightarrow{\text{lipase}} \text{-[ORC(=O)]}_n\text{-}$$

Scheme 14

Polycondensation of oxyacids or their esters

By Esterification (Dehydration). Dehydration polycondensation is the simplest polycondensation mode. The first paper appeared in 1985, reporting a lipase-catalyzed polycondensation of an oxyacid monomer, 10-hydroxydecanoic acid, in benzene using poly(ethylene glycol) (PEG)-modified lipase soluble in the medium. The degree of polymerization (DP) value of the product was more than 5. The PEG-modified esterase also induced the oligomerization of glycolic acid, the shortest oxyacid [39]. A lipase-catalyzed polymerization of lactic acid gave a low molecular weight poly(lactic acid) ($R = CH_3$, PLA) under a variety of the reaction conditions (Scheme 15) [40].

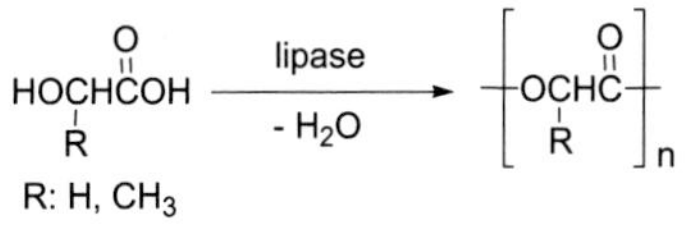

$$\underset{R}{\text{HOCHCOH}} \xrightarrow[-\text{H}_2\text{O}]{\text{lipase}} \text{-[OCHC(=O)]}_n\text{-}$$

R: H, CH_3

Scheme 15

Dehydration polycondensation of ricinoleic acid, a main component derived from castor oil, proceeded using lipases as catalyst at 35°C in a hydrocarbon solvent to give the polymer with molecular weight around 1×10^3. Ricinoleic acid was also polymerized via dehydration with immobilized lipase PC (*Pseudomonas cepacia*) catalyst to give the polymer with M_w up to 8,500 [41].

In the case of 11-hydroxyundecanoic acid, the polymer with relatively high molecular weight of 2.2×10^4 was obtained in the presence of activated molecular sieves [42]. Immobilized lipase from *Candida antarctica* lipase B (CALB, Novozym 435) was an efficient catalyst for the dehydration polycondensation of an oxyacid, cis-9,10-epoxy-18-hydroxyoctadecanoic acid isolated from the outer birch bark (Betula verrucosa), perfomed in toluene in the presence of molecular sieves at 75°C to give the polyester with the highest M_w 2.0 × 10^4 (M_w/M_n = 2.2) after 68 h [43].

Cpolymerization of 12-hydroxydodecanoic acid with methyl 12-hydroxystearate (both from seed oil) was catalyzed with Novozym 435 in toluene in the presence of molecular sieves at 90 °C. The feed ratio of the monomers and reaction time were varied. During the reaction water as well as methanol liberated. After several days, the copolymer was obtained in good yields, having M_w ~1.0 × 10^5 showing elasticity and biodegradability [44].

By Transesterification. Polycondensation of methyl ricinoleate by immobilized lipase PC catalyst produced poly(ricinoleate) with high molecular weight (M_w > 1 × 10^5) in bulk in the presence of molecular sieves at 80°C after 7 days. The polymer was a viscous liquid at room temperature with T_g of −74.8°C and biodegradable, which was readily cured to give chloroform insoluble crosslinked materials [45].

A regioselective polycondensation of isopropyl aleuriteate was achieved by Novozym 435 catalyst, where the only primary alcohol was reacted at 90°C (Scheme 16). The polymer of M_n 5,600 was obtained in 43% yields. Copolymerization of isopropyl aleuriteate with ε-CL gave a random copolymer having M_n up to 1.06 × 10^4 in ~70% yields [46].

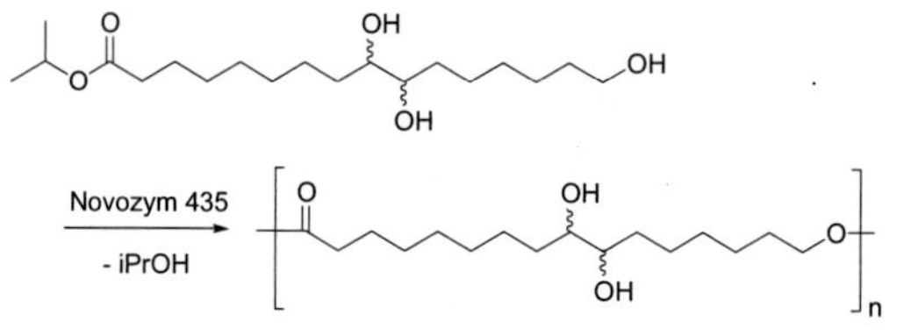

Scheme 16

Optically active oligoester was enantioselectively prepared from racemic 10-hydroxyundecanoic acid. The resulting oligomer was enriched in the (S) enantiomer with 60% enantio-excess (ee) and the residual monomer was recovered with 33% ee favoring (R) enantiomer [47].

Transesterification polycondensation of racemic AB type monomers having a secondary hydroxy group amd a methyl ester moiety led to chiral polyesters by iterative tandem catalysis. The concurrent actions of an enantioselective acylation catalyst (Novozym 435) and a racemization catalyst (Ru(Shvo)) brought about the high conversion of the racemic monomers to enantio-enriched polymers. AB type monomers used were typically methyl 6-hydroxyheptanoate, methyl 7-hydroxyoctanoate, methyl 8-hydroxynonanoate, and methyl 13-hydroxytetradecanoate. The polycondensation at 70°C in toluene gave a polyester in high yields having M_n around several thousands with ee higher than 74% [48].

Oleic acid-based polyester was prepared via chemoenzymatic method. first, oleic acid (a C-18 mono-ene acid from vegetable oils) was epoxidized by lipase catalyst with H_2O_2 oxidant, and then the intermolecular ring-opening addition between the epoxide group and COOH group thermally took place to produce the poly(oleic acid)-based polyester, which was further crosslinked by a diisocyanate compound to give a biodegradable material [49].

Polycondensation of carboxylic acids or their esters with alcohols

By Esterification (Dehydration). The dehydration polycondensation mode is given in Scheme 17.

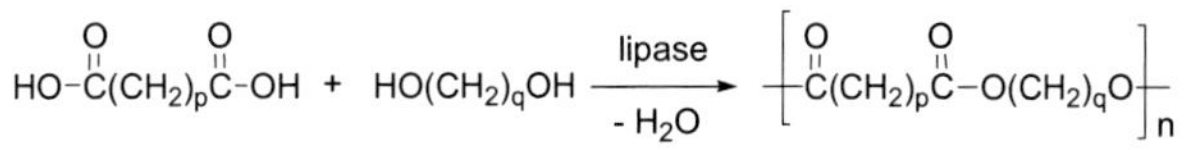

$$HO-\overset{O}{\underset{\|}{C}}(CH_2)_p\overset{O}{\underset{\|}{C}}-OH + HO(CH_2)_qOH \xrightarrow[-H_2O]{lipase} \left[\overset{O}{\underset{\|}{C}}(CH_2)_p\overset{O}{\underset{\|}{C}}-O(CH_2)_qO\right]_n$$

Scheme 17

In 1984, the first paper on the lipase-catalyzed polymerization reported an *Aspergillus niger* lipase (lipase A)-catalyzed dehydration polycondensation of several free dicarboxylic acids in the presence of an excess amount of a diol, to give oligomeric polyesters [50]. Higher molecular weight polyesters were enzymatically obtained by polycondensation of sebacic acid and 1,4-butanediol (p = 8, q = 4 in Scheme 17) under vacuum. By the *Mucor miehei* lipase (lipase MM)-catalyzed polymerization in hydrophobic solvents with high boiling point, the molecular weight of polyesters from various combinations of diacids and glycols reached the value $> 4 \times 10^4$ [51].

In the dehydration polycondensation, a high boiling diphenyl ether was a preferred solvent for the polycondensation of adipic acid (p = 4) and 1,8-octanediol (q = 8) giving M_n of 28 500 (48 h, 70°C). The reactions involving monomers having longer alkylene chain length of diacids (sebacic and adipic acids) and diols (1,8-octanediol and 1,6-hexanediol) showed a higher reactivity than the reactions of shorter chain length diacids and 1,4-butanediol [52]. In ionic liquids, a dehydration polycondensation gave effectively the product polyester with lipase catalysis [53].

Without using solvent, lipase CA (CALB) efficiently catalyzed the polycondensation of dicarboxylic acids and glycols under mild reaction conditions at 60°C, despite the initial heterogeneous mixture of the monomers and catalyst. Methylene chain length of the monomers greatly affected the polymer yield and molecular weight. The polymer with molecular weight higher than 1×10^4 was obtained by the reaction under reduced pressure [54]. CALB was covalently immobilized onto epoxy-activated macroporous poly(methyl methacrylate) Amberzyme beads and epoxy-activated nanoparticles (nanoPSG) with a poly(glycidyl methacrylate) outer region. In bulk Amberzyme-CALB catalyzed polycondensation between glycerol (0.1 equiv), 1,8-octanediol (0.4 equiv), and adipic acid (0.5 equiv) at 90°C for 24 h gave polyesters of M_w ~4.0×10^4 and the catalyst showed a better stability in recycled use over Novozym 435 [55].

A dehydration reaction is normally conducted in non-aqueous media. Since the product water of the dehydration is in equilibrium with starting materials, the solvent water disfavors the dehydration to proceed in an aqueous medium due to the "law of mass action". Nevertheless, lipase catalysis enabled a dehydration polycondensation of a dicarboxylic acid and a glycol in water at 45°C, to afford a polyester in good yields [56]. In the polymerization of an α,ω-dicarboxylic acid and a glycol, the polymerization behavior was greatly depending on the methylene chain length of the monomers. The polymer production needs appropriate hydrophobicity of monomers. This finding of dehydration in water is a new finding in organic chemistry.

Polyols including sugar components were used for dehydration polymerization, *e.g.*, the direct polycondensation between adipic acid and sorbitol in bulk was carried out with lipase CA catalyst (10 wt% relative to the monomers) at 90°C for 48 h. The product poly(sorbityl adipate) was water soluble. The M_n and M_w values were 10 880 and 17 030, respectively. Sorbitol was esterified at primary alcohol group of 1- and 6-positions with high regioselectivity (85 ± 5%). To obtain a water insoluble sorbitol copolyester, adipic acid, 1,8-octanediol, and sorbitol (molar ratio 50 : 35 : 15) were ter-polymerized at 90°C for 42 h (Scheme 18(a)). The methanol-insoluble part (80%) had M_w of 1.17×10^5. Another terpolymerization of adipic acid, 1,8-octanediol, and glycerol, in the ratio 50 to 40 to 10 mol%, was performed in bulk at 70°C, to give a polyester (Scheme 18(b)). The product polymer showed the monomers ratio in the 50 : 41 : 9, respectively, and the values for M_w and M_w/M_n of 75 600 and 3.1, respectively [57]. A similar dehydaration polycondensation to produce terpolymers was conducted by using monomers, adipic acid (A_2), 1,8-octanediol (B_2), and trimethyrolpropane (TMC, B_3) with lipase CA catalyst in bulk at 70°C for 42 h. An example of hyperbranched copolyester with 53% TMC adipate units was obtained in 80% yields, with M_w 14 100 (M_w/M_n = 5.3) [58].

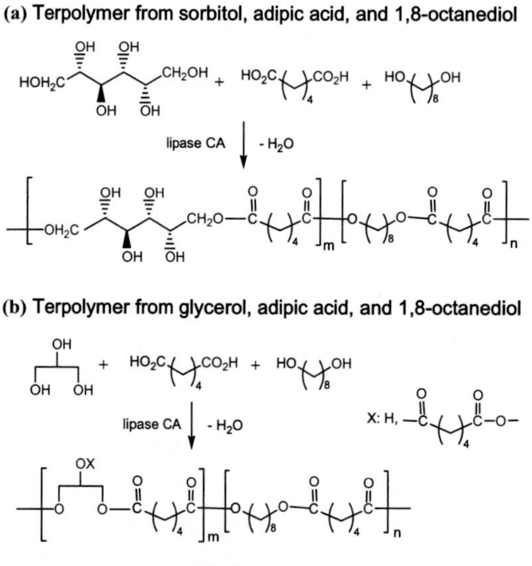

Scheme 18

Cutinase enzyme catalyzed a dehydration polycondensation [59]. A glycol and a diacid were combined for polycodensation with 1% w/w enzyme at 70°C for 48 h under vacuum. With fixing the adipic acid component, polyesters from 1,4-butanediol, 1,6-hexanediol, and 1,8-octanediol showed M_n values of 2,700, 7,000, and 12 000, respectively. With fixing the 1,4-cyclohexanedimethanol component, polyesters from succinic acid, adipic acid, suberic acid, and sebacic acid possessed M_n values of 900, 4,000, 5,000, and 19 000 for these C4, C6, C8, and C10 diacids, respectively.

There was a tendency that the higher hydrophobicity monomers gave the product polyester with the higher molecular weight.

In addition, the reaction of a linear polyanhydride like poly(azelaic anhydride) and a glycol like 1,8-octanediol was induced with lipase CA catalyst involving the dehydration to give a polyester with molecular weight of several thousands [60].

By Transesterification. Transesterification polycondensation normally needs activation of carboxylic acid groups, ordinarily by esterification of the acid group. In the early studies alkyl or haloalkyl esters and later vinyl esters have been often used.

A lipase-catalyzed high enantioselective polymerization was reported in 1989; the reaction of bis(2,2,2-trichloroethyl) *trans*-3,4-epoxyadipate with 1,4-butanediol in anhydrous diethyl ether using porcine pancreas lipase (PPL) catalyst gave a highly optically active polyester (Scheme 19). The feed molar ratio of the diester to the diol was adjusted to 2:1 so as to produce the (−)-polymer resulting in enantiomeric purity > 96%. The molecular weight was estimated as 5,300. The unchanged (+)-monomer was shown to have an enantiomeric purity higher than 95% [61].

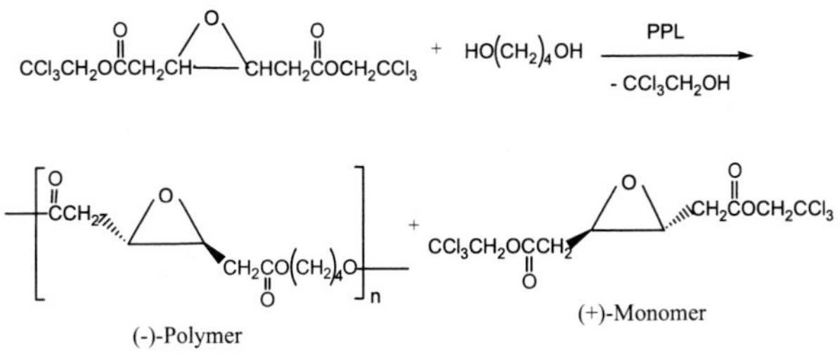

Scheme 19

Polycondensation between dimethyl succinate and 1,6-hexanediol catalyzed by lipase CA in toluene at 60 °C reached the ring-chain (cyclic-linear structure) equilibrium of the product polymer. Adsorption of methanol by molecular sieves or elimination of methanol by nitrogen bubbling shifted to the thermodynamic equilibrium. Polyesters with the molecular weight about several thousands were prepared from α,ω-alkylene dicarboxylic acid dialkyl esters and whatever the monomer structure, cyclic oligomers were formed (Scheme 20) [62].

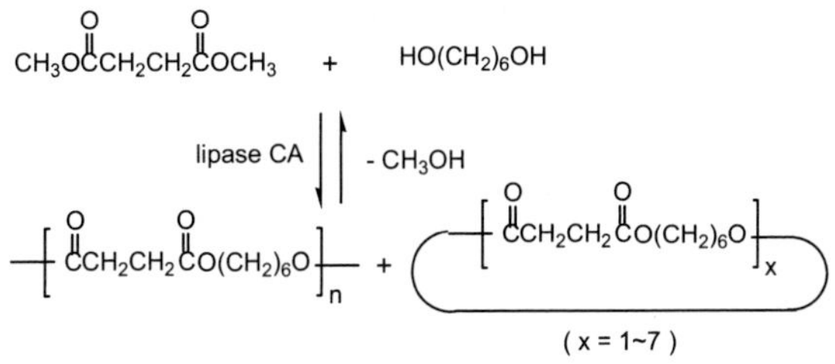

Scheme 20

Novozym 435-catalyzed synthesis of poly(butylene succinate) (PBS) via polycondensation was achieved using a monophasic reaction mixture of dimethyl succinate and 1,4-butanediol in bulk and in solution. Diphenyl ether was a preferred solvent to give a higher molecular weight PBS; at 60, 70, 80, and 90°C after 24 h, M_n values of PBS were 2,000, 4,000, 8,000, and 7,000, respectively. The reaction at 95°C after 21 h gave PBS with M_n value of 38 000 [63].

Green solvents, ionic liquids, were found to act as a good medium. The lipase CA-catalyzed polycondensation of diethyl adipate with 1,4-butanediol efficiently proceeded in 1-butyl-3-methylimidazolinium tetrafluoroborate or hexafluorophosphate under reduced pressure at 60°C to produce the polyester with M_n up to 1,500 after 72 h in 91% yields [53]. Polycondensation of diethyl sebacate or dimethyl adipate and 1,4-butanediol in 1-butyl-3-methylimidazolinium hexafluorophosphate took place at 70°C or even at room temperature [64].

A new method of dynamic kinetic resolution (DKR) was used to synthesize an optically active polyester from a racemic monomer. A mixture of stereoisomers of a secondary diol, α,α'-dimethyl-1,4-benzenedimethanol, was enzymatically polymerized with dimethyl adipate (Scheme 21) [65]. Due to the enantioselectivity of lipase CA, only the hydroxyl groups at the *(R)* center are preferentially reacted to form the ester bond with liberation of methanol. The reactivity ratio was estimated as *(R)/(S)* = ~1 × 10^{-6}. In situ racemization from the *(S)* to the *(R)* configuration by Ru catalysis allowed the polymerization to high conversion, that is, the enzymatic polymerization and the metal-catalyzed racemization occured concurrently. The DKR polymerization was carried out for 4 days; during the reaction molecular weight increased to 3,000–4,000 and the optical rotation of the reaction mixture increased from −0.6° to 128°.

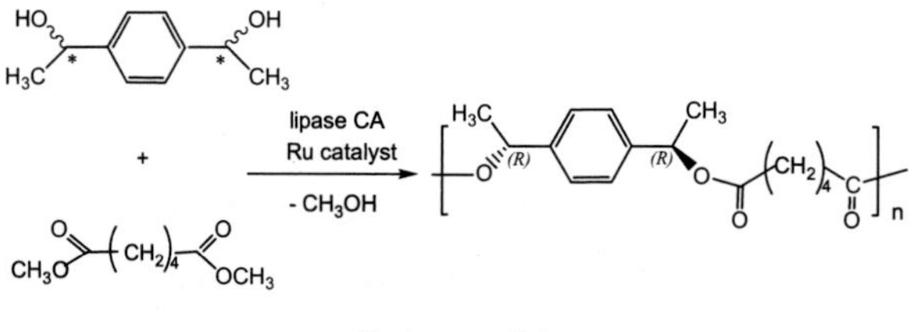

Scheme 21

Transesterifications using lipase catalyst are often very slow owing to the reversible nature of the reactions. To shift the equilibrium toward the product polymer more effectively, activation of esters was conducted by using a halogenated alcohol like 2-chloroethanol, 2,2,2-trifluoroethanol, and 2,2,2-trichloroethanol. Compared with methanol or ethanol, they increased the electrophilicity of the acyl carbonyl and avoided significant alcoholysis of the products by decreasing the nucleophilicity of the leaving alkoxy group.

In 1994, an irreversible process was developed by using a vinyl ester for the lipase-catalyzed acylation, where the product of vinyl alcohol tautomerizes to acetaldehyde [66]. The reaction of an alcohol with a vinyl ester proceeds much faster than with an alkyl ester or a haloalkyl ester to form the desired product in higher yields. A divinyl ester was employed for the first time as the activated acid form in the enzyme-catalyzed polyester synthesis. The lipase PF-catalyzed polycondensation of divinyl adipate and 1,4-butanediol was performed at 45°C in diisopropyl ether for 48 h to afford a polyester with M_n of 6.7×10^3 (Scheme 22), whereas the use of adipic acid and diethyl adipate did not produce the polymeric materials under the similar reaction conditions. As a diol, ethylene glycol, 1,6-hexenediol, and 1,10-decanediol were also reacted to give the corresponding polyester with molecular weight of several thousands. The same polycondensation of divinyl adipate and 1,4-butanediol with lipase PC (*Pseudomonas cepacia*) catalyst produced the polyester with M_n of 2.1×10^4 [67].

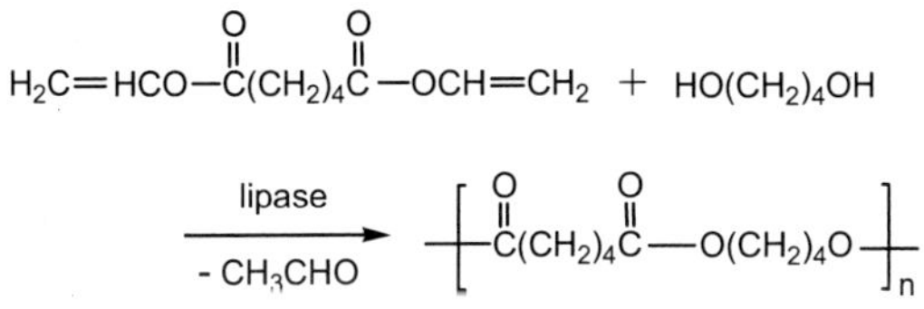

$$H_2C=HCO-\overset{O}{\overset{\|}{C}}(CH_2)_4\overset{O}{\overset{\|}{C}}-OCH=CH_2 + HO(CH_2)_4OH$$

$$\xrightarrow[-CH_3CHO]{\text{lipase}} \left[\overset{O}{\overset{\|}{C}}(CH_2)_4\overset{O}{\overset{\|}{C}}-O(CH_2)_4O\right]_n$$

Scheme 22

Supercritical carbon dioxide (scCO$_2$) was shown to be a good solvent for the lipase-catalyzed polycondensation of divinyl adipate and 1,4-butanediol. Quantitative consumption of both monomers was achieved to give the polyester with M_n of 3.9×10^3 [68].

In an ionic liquid 1-butyl-3-methy-imidazolium tetrafluoroborate ([bmim][BF$_4$]), a similar polycondensation between diethyl adipate or diethyl sebacate and 1,4-butanediol gave the polyester having M_n up to 1,500 in good yields. Since the ionic liquid is non-volatile, ethanol was removed under vacuum during the reaction [53]. Lipase CA-catalyzed polycondensation of dimethyl adipate or dimethyl sebacate with 1,4-butanediol was performed in an ionic liquid such as [bmim][BF$_4$], [bmim][PF$_6$], and [bmim][(CF$_3$SO$_2$)$_2$N] at 70°C for 24 h to give a higher molecular weight polyester, M_n reaching several thousands. Using ionic liquids as solvent involves the wide range of tunablity of solvent hydrophilicity and monomer solubility for the reaction [64].

Aromatic polyesters were produced efficiently by the lipase CA-catalyzed polycodensation of aromatic diacid divinyl esters. Divinyl esters of isophthalic acid, terephthalic acid, and *p*-phenylene diacetic acid were polymerized with various glycols to give polyesters containing aromatic moiety in the main chain with the highest M_n of 7,200 in heptane at 60°C for 48 h [69].

Polycondensation of polyols was achieved regioselctively with lipase catalysis by using divinyl esters. Triols such as glycerol were regioselectively polymerized at a primary hydroxyl group with divinyl adipate by lipase catalyst to produce a linear polyester linked through mainly primary hydroxyl group having also a secondary hydroxyl group (5–10%) in the main chain. The polymer contained pendant hydroxyl groups, with no evidence of network structure, having M_w value from ~3,000 to 14 000 [70]. The reaction of divinyl sebacate and glycerol with lipase CA catalyst produced water-soluble polyesters with M_w up to 2,700 in bulk at 60°C for 8 h under argon. The chloroform-soluble part with M_w of 19 000 was isolated in 63%, which indicated the regioselectivity of primary OH/ secondary OH ratio of 74/26. At 45°C, however, the regioselectivity was perfectly controlled to give a linear polymer consisting exclusively of 1,3-glyceride unit [71]. The lipase CA catalysis gave a reduced sugar-containing polyester regioselectively from divinyl sebacate and sorbitol, in which sorbitol was exclusively acylated at the primary alcohol of 1- and 6-positions in acetonitrile at 60°C for 72 h (Scheme 23). Mannitol and *meso*-erythritol were also regioselectively polymerized with divinyl sebacate [72].

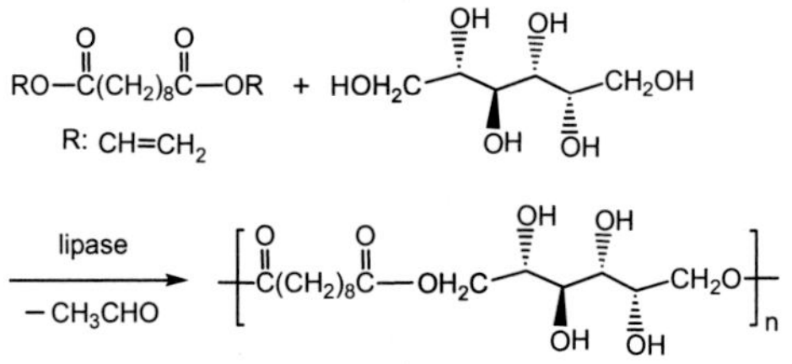

Scheme 23

Lipase-catalyzed terpolymerization of divinyl esters, glycols, and lactones produced ester terpolymers with M_n higher than 1×10^4. Lipases showed high catalytic activity for the terpolymerization involving both polycondensation and ring-opening polymerization manners simultaneously in one-pot to produce ester terpolymers, without involving homo-polymer formation [73]. A similar terpolymerization was performed using three kinds of monomers, ω-pentadecalactone, diethyl succinate, and 1,4-butanediol, by CALB catalyst desirably at 95°C via a two-stage vacuum technique. The polymerization was examined under various reaction conditions and the product terpolyester reached M_w 77 000 with $M_w/M_n \sim 1.7$–4.0 [74].

These results accord with the frequent occurrence of an intermolecular transesterification between the resulting polyesters during the polymerization; from a mixture of two homo-polyesters a copolyester was obtained by lipase catalysis [75].

New crosslinkable polyesters were prepared by the lipase-catalyzed polycondensation of divinyl sebacate with glycerol in the presence of an unsaturated higher fatty acid (i) such as linoleic acid and linolenic acid obtained from renewable plant oils (route A, Scheme 24). Product polyester (iii) is biodegradable and contains an unsaturated fatty acid moiety in the side chain. Curing of (iii) proceeded by oxidation with cobalt naphthenate catalyst or thermal treatment gave a crosslinked transparent film. Biodegradability of the film obtained was verified by biochemical oxygen demand (BOD) measurement [76]. Furthermore, epoxide-containing polyesters were enzymatically synthesized via two routes using unsaturated fatty acids (routes A and B, Scheme 24). In route A, (iii) was enzymatically epoxidized to give (iv), and in route B the epoxidization of the fatty acid was first conducted and lipase-catalyzed polycondensation of the product (ii) was performed to produce (iv) [77]. Curing of (iv) proceeded thermally, yielding transparent polymeric films with high gloss surface. Pencil scratch hardness of the film was enhanced, compared with that of the cured film from (iii). The obtained film showed good biodegradability in the activated sludge test.

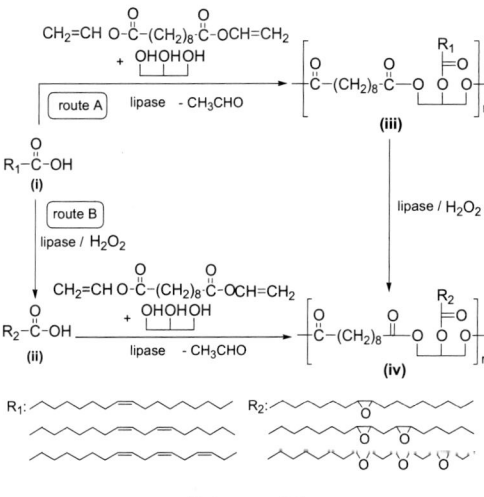

Scheme 24

Another cross-linkable group can be a mercapto group. Direct lipase CA-catalyzed polycondensation of 1,6-hexanediol and dimethyl 2-mercaptosuccinate at 70°C in bulk gave an aliphatic polyester having free pendant mercapto groups with M_w = 14 000 in good yields. The mercapto group content could be controlled by copolymerization with other monomers. The polyester was readily cross-linked by the air-oxidation via the disulfide linkage formation in dimethyl sulfoxide [78].

Ring-opening polymerization

Ring-opening polymerization (ROP) has been most extensively studied for the polyester synthesis. Typical examples of cyclic monomers studied by an enzyme catalyst are shown in Figure 7.

ROP of Cyclic Esters (Lactones). Catalytic activity for ROP of lactone monomers have been examined on a variety of lipase from different origin such as *Pseudomonas fluorescens* (lipase PF), *Candida cylindracea* (lipase CC), porcine pancreas lipase (PPL), *Aspergillus niger* (lipase A), *Candida rugosa* (lipase CR), *Penicillium roqueforti* (lipase PR), *Pseudomonas cepacia* (lipase PC), *Mucor miehei* (lipase MM), and *Rhizopus japanicus* (lipase RJ). Among others, *Candida antarctica* (lipase CA, CALB, or Novozym 435) has been most often used due to the increased activity recently [2].

In 1993 lipase-catalyzed ROP was first found for ε-caprolactone (ε-CL, 7-membered) (scheme 25(a)) and δ-valerolactone (δ-VL, 6-membered) by two independent groups [79,80]. ROP of ε-CL by lipase PF (*Pseudomonas fluorescens*) in bulk at 75°C for 10 days gave poly(ε-CL) in 92% yields, having molecular weight M_n of 7,700 with M_w/M_n = 2.4. Similarly, δ-VL yielded at 60°C poly(δ-VL) having M_n 1,900 with M_w/M_n = 3.0 [80a]. These polyesters possessed the terminal structure of a carboxylic acid group at one end and a hydroxyl group at the other, indicating that

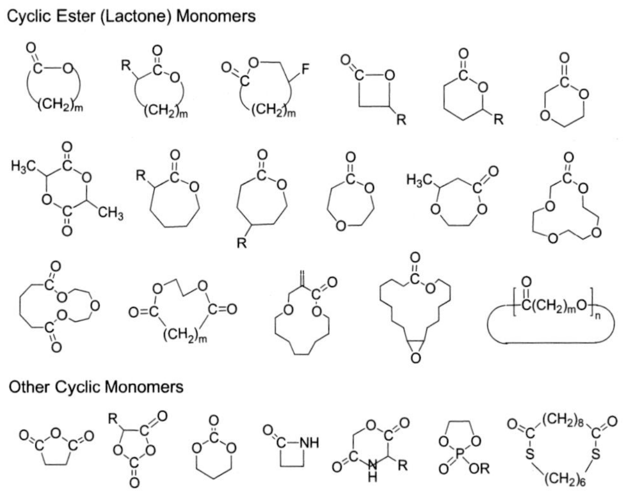

Figure 7 Examples of cyclic monomers for enzyme-catalyzed polyester synthesis. (Reprinted with permission from ref 2e. Copyright 2010 The Japan Academy.)

the ROP was initiated by water molecule and terminated also by water, when nothing was added. Catalyst activity was examined with using the ROP of ε-CL for lipases from *Candida antarctica* B (CALB), *Rhizomucor meihei* (lipase RM), *Candida rugosa* (lipase CR), and lipase PF, in the supported form or free form [80b].

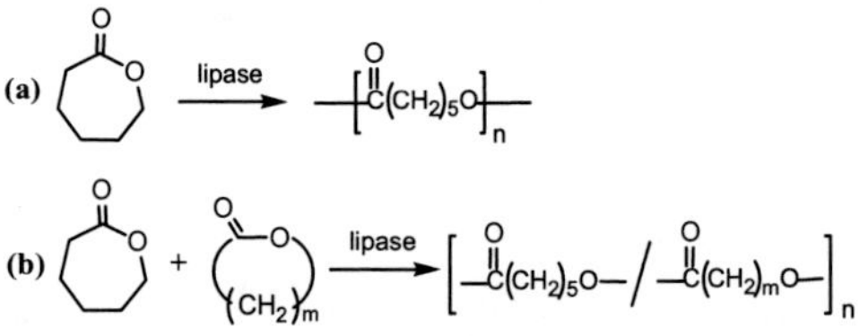

Scheme 25

Enzymatic ring-opening copolymerization by lipase PF catalyst between ε-CL and δ-VL was achieved in bulk at 60°C for 10 days, giving rise to a copolyester of M_n value 3,700 with M_w/M_n = 2.9, with a random copolyester structure (Scheme 25(b), m = 4). Copolymerization of ε-CL with other lactones like 15-pentadecanolactone (PDL) (m = 14 in Scheme 26) and D-lactide was also achieved [80b]. By using PPL as catalyst and methanol as initiator, ε-CL was polymerized in hexane at 45°C for up to 26 days to complete monomer conversion, affording poly(ε-CL) and dilactone [79]. Following these works, various lactones (cyclic esters) of different ring size, unsubstituted and substituted, and also other cyclic monomers, have been extensively studied for ring-opening polymerization and copolymerization [2,5,81].

Scheme 26 shows unsubstituted 4- to 17-membered lactones so far polymerized by lipase catalyst for ROP. All monomers showed a good ROP reactivity catalyzed by lipase. The polymerization can be performed in bulk, in an organic solvent or in other solvents.

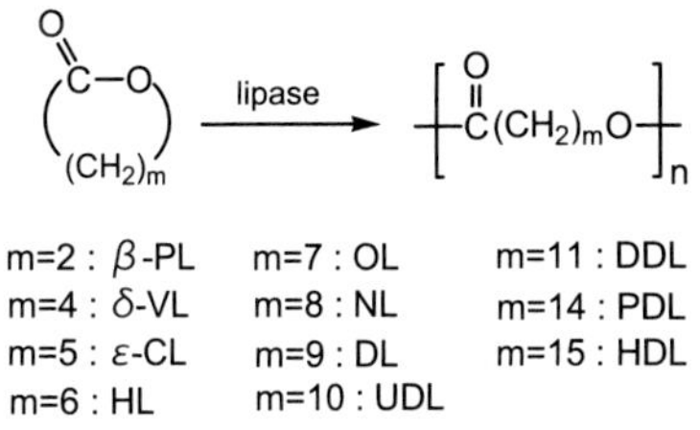

m=2 : β-PL   m=7 : OL   m=11 : DDL
m=4 : δ-VL   m=8 : NL   m=14 : PDL
m=5 : ε-CL   m=9 : DL   m=15 : HDL
m=6 : HL   m=10 : UDL

Scheme 26

A 4-membered lactone (β-propiolactone, β-PL) and substituted β-PLs were polymerized with lipase catalyst in bulk in 1996 by four groups, yielding linear polymers with molecular weight up to 2×10^4 [82] and also cyclic oligomers [82b]. Propyl malolactonare (β-propyloxycarbonyl-β-PL) was polymerized with lipase CR catalyst via ROP in toluene and in bulk. The highest reaction rate was achieved in toluene with 2.11 M monomer at 60°C at 10 wt% enzyme amount for the monomer, which was 25 times faster than the thermal polymerization observed. The maximum M_n value of 5,000 of the polyester was obtained almost quantitatively. Thermal ROP gave the polymer showed M_n of 1,800 [83].

Lactide, a six-membered cyclic dimer of lactic acid, is a very important starting monomer for the production of poly(lactic acid) (PLA) as green plastics and other application materials. In 1997, lipase PS-catalyzed ROP of lactide was reported, where the ROP was carried out in bulk at a temperature between 80 and 130°C to produce PLA with M_w up to 1.26×10^5, but the product yield was low. The DL-LA gave the higher molecular weight in comparison with LL-LA (LLA) and DD-LA (DLA) monomers [84]. A recent paper revealed that Novozym 435-catalyzed ROP of LLA did not take place, whereas that of DLA was enantioselectively induced to produce polyDLA of M_n value 3,300 (M_w/M_n = 1.2) in 33 % yields. The polymerization employed the catalyst amount of 12.5% for DLA in toluene at a lower reaction temperature of 70°C for 3 days (Scheme 27) [85].

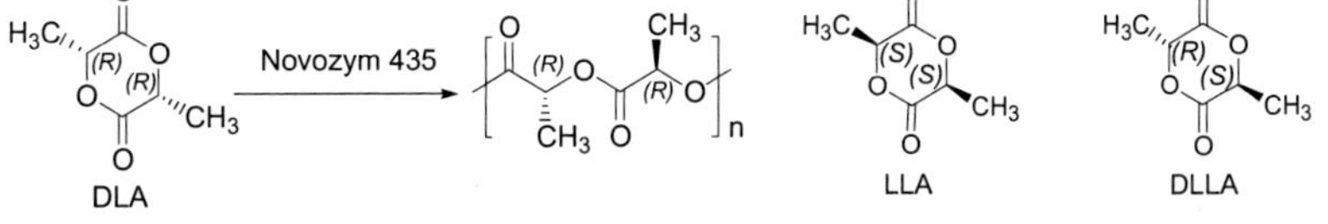

Scheme 27

A 7-membered lactone of ε-CL is the most extensively studied among various lactone monomers (m = 5 in Scheme 26) [2c,5,8,86,87]. Lipase CA was found as the most effective for the polymerization of ε-CL [88]. For example, the catalyst amount was reduced to 1%, which is compared with the reported systems of 20 to 50% of lipase from other origins. The reaction time could be reduced from 10 days to less than 10 h in bulk at 60°C. The high catalytic activity was observed also for 12- and 13-membered lactones (UDL and DDL, respectively). The molecular weight of poly(ε-CL) of 25 000 was readily reached by ROP in toluene at 70°C. Multifunctional initiators from polyglycidols having primary OH groups were employed for initiating the ROP of ε-CL by enzymatic and chemical catalysts. Novozym 435 and Zn(II) 2-ethylhexanoate were used for the enzymatic ROP and for the chemical catalyst, respectively. The obtained polymer architectures were of core-shell polymers and the difference in structure was discussed for polymers produced via enzymatic and chemical processes [89].

Oligomers from ε-CL gave poly(ε-CL) with M_n greater than 7.0×10^4, when the reaction was carried out at 70°C in toluene [87d]. Cyclic dimer of ε-CL (14-memberd) was polymerized by lipase CA, affording quantitatively poly(ε-CL) with M_n of 89 000 at a higher reaction temperature, and also another large cyclic oligomer was polymerized [90].

The lipase-catalyzed ROP reactivity of methyl substituted ε-CL monomers was examined, and it was revealed that ω-methyl ε-CL showed the least polymerizability among unsubstituted, α-and γ-substituted ε-CL monomers [87j]. By using lipase CA, α-methyl ε-CL produced the corresponding polyester with M_n 11 400 (M_w/M_n = 1.9) in 74 % yields in bulk at 60°C for 24 h and α-methyl δ-VL afforded the polyester (M_n 8,400; M_w/M_n = 2.0) in 93% yields at 45°C [87k].

Cutinase from *Humicola insolens* (HiC) showed a high catalytic activity for ROP of ε-CL and PDL, for examples, the reaction of ε-CL with 0.1% w/w HiC at 70°C in bulk for 24 h gave the polyester of M_n = 16 000 (M_w/M_n = 3.1) in > 99% yields, and monomer PDL with the same catalyst amount produced polyPDL of M_n = 44 600 (M_w/M_n = 1.7) in toluene at 70°C also in > 99% yields [59].

A 9-membered lactone (8-octanolide, OL) was also polymerized by lipase catalyst, producing the polymer with molecular weight of 1.6×10^4 at 75°C after 10 days [91]. As to macrolides, 11-undecanolide (12-membered, UDL), 12-dodecanolide (13-membered, DDL), 15-pentadecanolide (16-membered, PDL), and 16-hexadecanolide (17-membered, HDL), were enzymatically polymerized [86,92]. At 75°C for 10 days in bulk, polyUDL of M_n = 23 000 (M_w/M_n = 2.6) was obtained quantitatively by lipase PF and the same polyester M_n = 25 000 (M_w/M_n = 2.2) in 95% yields by lipase CC, which suggests a much higher ring-opening polymerizability of the macrolide than ε-CL [91].

Ring-opening copolymerization of PDL with four monomers, δ-VL, ε-CL, UDL, and DDL, by lipase PF and lipase PC at 60°C in bulk was carried out to give copolymers with M_n ranging 1,200–6,300, being not statistically random [86a]. ROP of PDL was investigated in more detail; at a low reaction water level, lipase I-PS-30 catalyzed polymerization of PDL at 70°C gave polyPDL of M_n and M_w/M_n of 62 000 and 1.9, respectively [92c]. Lipase CA (CALB) catalyzed the ring-opening copolymerization of PDL with 1,4-dioxan-2-one (DO) in toluene or diphenyl ether at 70°C for 26 h gave a copolyester poly(PDL-*co*-DO) with high M_w (> 30 000). A larger ring monomer PDL showed a higher polymerizability than a smaller counterpart DO [93].

The largest lactone monomer having a simple unsubstituted structure so far studied is HDL

(17-membered). ROP of HDL was performed by lipase CA, lipase CC, lipase PC, lipase PF, or PPL in bulk at 45–75°C for 5 days, giving rise to polyHDL with M_n reaching to 5,800 (M_w/M_n = 2.0) in good or quantitative yields [92d]. A 26-membered sophorolipid lactone having a double bond was prepared and subjected to ring-opening metathesis polymerization with a Ru catalyst, giving rise to poly(sophorolipids), with M_n exceeding 1.0×10^5 [94].

Reaction Solvents. Lipase-catalyzed ROP is normally carried out in bulk or in an organic solvent like toluene, heptane, 1,4-dioxane, diisopropyl ether, and dibutyl ether. In addition, water, supercritical carbon dioxide (scCO$_2$) and ionic liquids could be employed as solvent, which are regarded as typical "green solvents".

The first example using water as solvent used five lactone monomers, ε-CL, OL, UDL, DDL and PDL for the lipase-catalyzed ROP (Scheme 26). Macrolides of UDL, DDL and PDL were polymerized by lipase in water to produce the corresponding polyesters up to 89% yields. *Pseudomonas cepacia* (lipase PC) showed the best results in terms of polyester yields and molecular weight [86c,95]. The second example of the water medium system is the lipase-catalyzed ROP of a lactone in miniemulsions. PolyPDL nanoparticles were prepared, which is considered to be a direct synthesis of biodegradable polymer nonoparticles (size < 100 nm). PolyPDL showed a bimodal molecular weight distribution; the majority was that of high molelcular weight ($> 2.0 \times 10^5$) [96].

Supercritical carbon dioxide (scCO$_2$) is inexpensive, inert, nontoxic, and nonflammable, and possesses potentials for polymer synthesis and recycling. scCO$_2$ was used for the first time for the lipase CA-catalyzed ROP of ε-CL to produce poly(ε-CL) with molecular weight M_w ~1.1×10^4 in high yields and for the ring-opening copolymerization between ε-CL and DDL to produce poly(ε-CL-*co*-DDL) with M_w 1.3×10^4 [68]. A further work reported the synthesis of poly(ε-CL) having a higher molecular weight (M_w) reaching 7.4×10^4. The enzyme and scCO$_2$ were repeatedly used for the polymerization [97].

Ionic liquids are nonvolatile, thermally stable, and highly polar liquids, which allow to dissolve many organic, inorganic, metallo-organic compounds, and also polymeric materials. Thus, ionic liquids become popular for the synthesis and modification of polymers from the standpoint of green character. Lipase-catalyzed ROP of ε-CL was realized for the first time in an ionic liquid solvent, such as 1-butyl-3-methyl-imidazolium salts ([bmim][X$^-$]), giving rise to poly(ε-CL) with M_n 4,200 (M_w/M_n = 2.7) in 97% yields at 60°C [53]. Novozym 435-ctalyzed ROP of ε-CL in three ionic liquids, [bmim][BF$_4$], [bmim][PF$_6$], and [bmim][(CF$_3$SO$_2$)$_2$N] at 60°C for 24 h produced poly(ε-CL) of a higher M_n 7,000–9,500 (M_w/M_n ~2.4) in good yields [64].

End-Functionalized Polyesters. Synthesis of end-functionalized polymers, typically macromonomers and telechelics, is both fundamentally and practically important in polymer chemistry, which requires a precise control of the polymer terminal structure. Lipase catalysis provides a novel method for a single-step synthesis of end-functionalized polyesters via a relatively simple reaction.

Initiator Method: Lipase catalysis involves a nucleophile like water and an alcohol to initiate the ROP of lactones. In fact, lipase CA-catalyzed ROP of ε-CL or DDL at 60°C in bulk in the presence of a functional alcohol produced end-functionalized polyesters ("initiator method") [98].

Functional alcohols include 2-hydroxyethyl methacrylate (HEMA), 5-hexen-1-ol and 5-hexyn-1-ol for the synthesis of methacryl-, ω-alkenyl- and alkynyl-type polyester macromonomers having M_n 1,000–3,100. The functionality reached 100% in many reaction runs (Scheme 28(a)). The lipase CA-catalyzed ROP of ε-CL or DXO initiated by 4-pentene-2-ol gave a macromonomer having a double bond. Macro-initiators, poly(ethylene glycol) and poly(ε-CL)-diol, initiated the ROP of ε-CL or 1,5-dioxepane-2-one (DXO) to give a triblock polyester with a hydroxyl group at both ends (Scheme 28(b)) [99]. The methacryl-type polyester macromonomer was derived via lipase-catalyzed ROP of PDL initiated with HEMA and radically polymerized to lead to polymers of a brush structure [100].

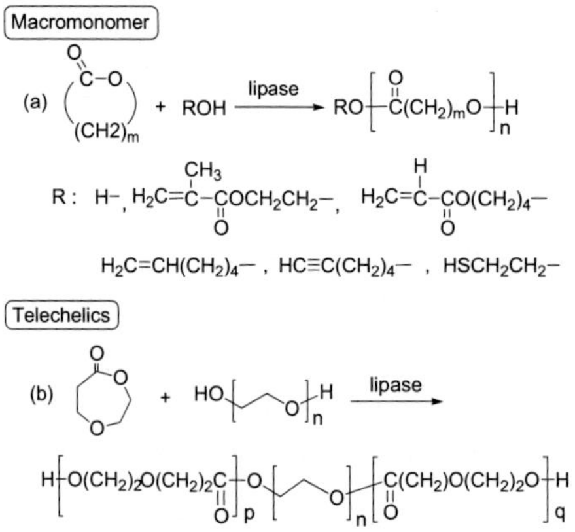

Scheme 28

A primary alcohol group in the 6-position of alkyl glucopyranosides, and six primary alcohol groups of a first generation dendrimer induced the ROP of ε-CL with lipase CA catalyst, where the regioselective initiation (monoacylation of the initiator) took place. Thus, the primary hydroxy group was regioselectively acylated to give the polymer bearing the sugar moiety at the polymer terminal [101]. The initiation process was examined under various reaction conditions at 60°C in bulk or in toluene, which relates to synthesize end-functionalized polymers by introducing the functionality into polymers via ROP (initiator method) [102].

A new approach to a biodegradable polyester system was performed by the lipase-catalyzed ROP of ε-CL and 1,4-dioxan-2-one monomers initiated from an alcohol attached on the gold surface. The polyester system can be used as a biocompatible/biodegradable polymer for coating materials in biomedical area such as passive or active coatings of stents [103].

The initiator method is useful for a single-step synthesis of end-functionalized polyesters as well as other polymers, which can be prepared via polymerization of a monomer induced by a lipase/functional alcohol (or another nucleophile) catalyst system.

Terminator Method: A single-step, convenient production of end-functionalized polyesters was developed by lipase-catalyzed ROP of DDL in the presence of vinyl esters [104]. The vinyl ester acted as terminator during the polymerization ("terminator method"). In use of vinyl (meth)-

acrylate as terminator, the (meth)acryl group was quantitatively introduced at the polymer terminal at 60°C in bulk to give a (meth)acryl-type polyester macromonomer; with M_n 2,000–4,000 and functionality > 0.95. The polymerization in the presence of vinyl 10-undecanoate produced the ω-alkenyl-type macromonomer (Scheme 29(a)). This system was applied to the synthesis of telechelics having a carboxylic acid group at both ends by using divinyl sebacate in the reaction mixture at 60°C in bulk; M_n of the telechelics was 2,900 with functionality of 1.95 (Scheme 29(b)) [105].

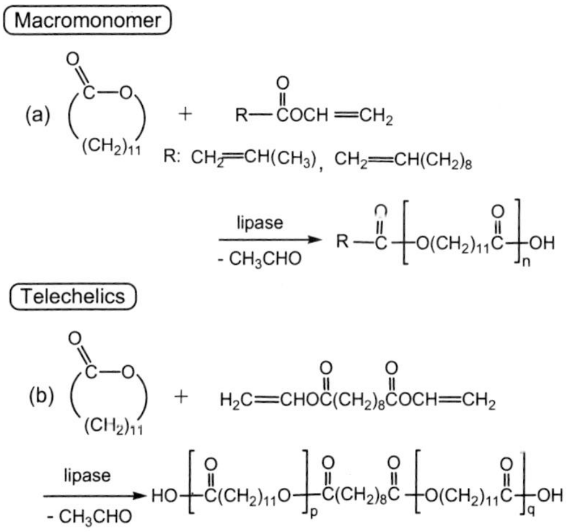

Scheme 29

An enzymatic one-pot synthesis gave difunctional telechelics. When the CALB-catalyzed ROP of PDL was conducted at 60°C by 6-mercaptohexanol initiator and then terminated with γ-thiobutyrolactone to give a telechelic polyester with very high content of thiol-thiol end-groups and with vinyl acrylate to produce that of thiol-acrylate end-groups [106]. A telechelics having thiol-thiol end-groups of polyPDL was synthesized by lipase-catalyzed ROP of PDL monomer with using 6-mercapto-1-haxanol as initiator and γ-thiobutyrolactone as terminator, and then used further for the preparation of semicrystalline polymer networks [107].

The terminator method involves a lipase-catalyzed single-step acylation of a polyester alcohol end-group with a vinyl ester or a lactone, the process of which can be applied basically for other polymer alcohols to give an end-functionalized polymer.

ROP Mechanism. Mechanism of lipase catalyzed ROP of lactones is described by considering the principal reaction course involving an acyl-enzyme intermediate (Scheme 30). Lipase-catalyzed hydrolysis of an ester *in vivo* is generally accepted to proceed via a similar acyl-enzyme intermediate. Catalytic site of lipase is known to be -CH$_2$OH group of a serine-residue. The key step is the reaction of lactone with lipase catalyst involving an enzyme-lactone complex and ring-opening of lactone to give an acyl-enzyme intermediate (enzyme-activated monomer, EM). This is an acylation step of the enzyme. The initiation is nucleophilic attack of a nucleophile such as water or an alcohol, which is present in the reaction mixture, at the acyl carbon of the intermediate to produce ω-hydroxycarboxylic acid (n = 1); the shortest propagating species. In the propagation stage, EM is nucleophilically attacked by the terminal hydroxyl group of a propagating chain end to produce a one-monomer-unit elongated polymer chain. Thus, the polymerization proceeds via an "activated monomer mechanism".

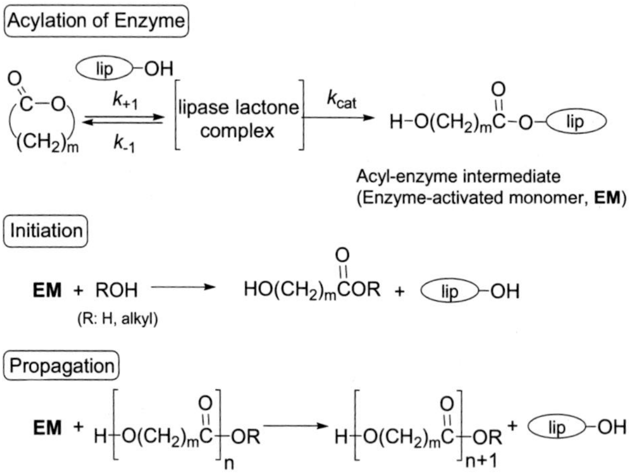

Scheme 30

ROP reactivity of monomers has been extensively studied for the unsubstututed monomers of different ring size shown in Scheme 26. The accumulated data show that the anionic ring-opening reactivity is governed primarily by the ring strain; the larger the ring strain, the higher the ROP reactivity of the monomer. Normally, a smaller ring-sized monomer showed a higher ROP reactivity. For example, in the Zn-catalyzed ROP, the reactivity ratio of the rate constant values for monomers (the ring-size in Scheme 26); 6: 7: 9: 12,13,16,17 = 2,500: 330: 21:1.0. The macrolide monomers (m = 10~15) indicated a similar small reactivity, whereas δ-VL (m = 4) had a 2,500 times higher reactivity [108].

In 1997, the first kinetic analysis based on the Michaelis-Menten equation was reported for the lipase PF (*Pseudomonas fluorescens* lipase)-catalyzed ROP of ε-CL and DDL [86b]. The lipase-catalyzed ROP obeyed the Michaelis-Menten kinetics. The formation of an acyl-enzyme intermediate (EM in Scheme 30) was considered to be the rate-determining of the overall reaction, because the reactivity of the intermediate must be high and hence the subsequent steps are rapid. The Michaelis-Menten kinetics parameters, K_m and V_{max} values, were determined and it was concluded quantitatively that DDL is 1.9 times more reactive than ε-CL in ROP [86c]. The V_{max}/K_m (s^{-1}) value is a good measure for the overall rate of the enzymatic ROP. Thus, the relative rate of lactone monomers with different ring size was derived with the ring-size; 6:7:12:13:16:17 = 0.1: 0.1: 0.13:0.19: 0.74: 1.0 [109]. With lipase PF, there is a tendency that the larger the ring-size, the larger the polymerization rate in one magnitude difference. that is, the higher polymerizability of the macrolides was explained by the higher rate in the formation of EM. As for the Michaelis-Menten constant K_m $((k_{-1}+ k_{cat})/k_{+1}$ mol·L^{-1}), the values are not much different each other; they are from 0.61 to 1.1, on the other hand, the maximal reaction rate V_{max} (k_{cat} [E]$_0$, mol·L^{-1}·s^{-1}) values are much more varied from 0.66 to 7.2. These results suggested that the reaction rate is mainly governed by the larger value of V_{max}, and much less by the binding ability. In other words, the reaction process from the lipase-lactone complex to form the intermediate EM is the key step in the lipase-catalyzed ROP.

In comparison with lipase PF, catalysis of lipase CA was found much different in behaviors. Lipase CA exhibited a higher catalytic activity for ε-CL than for macrolides. Qualitatively, initial rates (× 10^5, L·mol^{-1}·h^{-1}·mg^{-1}) of ROP by lipase CA catalyst were obtained as 2,300 for ε-CL, 48 for OL, 1,600 for DDL, and 4,700 for PDL. For these four monomers, the corresponding values of lipase PF were 1.3, 1.8, 2.5, and 8.5 [91]. Catalysis of lipase CA for ROP of nine lactones was investigated according to Michaelis-Menten kinetics [110]. The values of K_m are in a narrow range between 0.09 and 0.73 mol·L^{-1}, suggesting close affinities of the lipase for all lactones. On the other hand, the values of V_{max} varied between 0.07 and 6.10 mol·L^{-1}·h^{-1} and did not show a trend. These are alike partially to the results observed above for lipase PF catalysis [109], indicating that the ROP reactivity with lipase CA catalyst is likely operated also by the process from the lipase-lactone complex to form EM. The relative rate values look very complicated; there is no monotonous change depending on the ring size.

Chemoselective Polymerization. A chemoselective ROP of 2-methylene-4-oxa-12-dodecanolide, a macrolide derivative of methyl methacrylate, yielded a polyester having the reactive exo-methylene group in the main chain (Scheme 31). This type of polymer is hardly obtained by using a conventional chemical initiator. The lipase-catalyzed polymerization selectively afforded polyesters through the ring-opening process, whereas anionic and radical initiators induced the vinyl polymerization. The polyester was readily cross-linked radically via reaction through the reactive methacrylic methylene group to produce a cross-linked polymer [111].

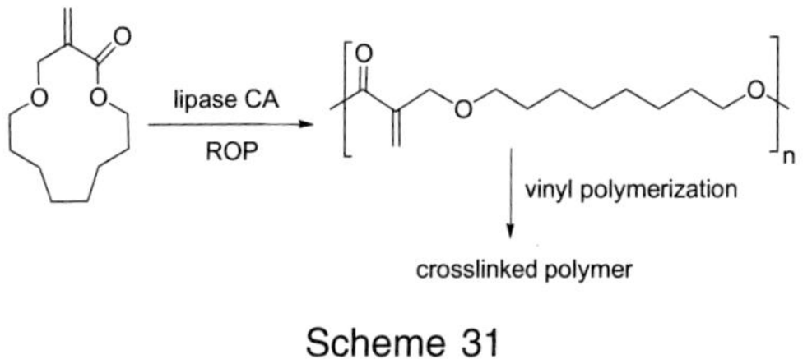

Scheme 31

A chemoselective ROP of Ambrettolide (Am) epoxide, a 17-membered macrolide having an epoxy group at 10-position (Figure 7), was performed by Novozym 435 catalyst to afford the polyester with M_n value of 9,700 (M_w/M_n = 1.9). The epoxide group remained unaffected during the polymerization [46]. Related functional macrolides, Globalide (Gl) and Am, are presently used in fragrance industry. Gl is a 16-membered lactone having the double bond at 11 or 12 position, and Am is a 17-membered lactone with the double bond at 10 position. Novozym 435-catalyzed

ROP of Gl and Am proceeded in toluene at 60°C in the presence of molecular sieves to give polyGl and polyAm, having M_n both around 2.4×10^4. Both polyesters had melting points 46–55 °C, which are compared with those of saturated polyesters, polyPDL and polyHDL, of around 95°C [112].

Regioselective Polymerization. ROP of ε-CL was initiated regioselectively from a sugar molecule with lipase-catalyst. The primary hydroxyl group in the 6-position of methyl or ethyl glucopyranoside induced the ROP of ε-CL in a high conversion of the monomer at 70°C. Lipase CA and PPL showed a high catalytic activity [101]. A sugar core containing methacryl-type macromonomer was also developed [113].

A one-pot, lipase-catalyzed, graft-polymerization of ε-CL and β-butyrolactone (β-BL) onto chitin and chitosan was accomplished in bulk at 70°C. The ROP of the cyclic monomers was initiated regioselectively from 6-OH group in chitin and from 6-OH and NH groups in chitosan to produce chitin-*graft*-polyester and chitosan-*graft*-polyester, respectively. In the products, both the stem polymer cellulose and the graft polyester are biodegradable [114].

A stereochemical factor of the substrate was controlled by lipase catalyst. Since benzyl alcohol is a good initiator, a backbone polymer, poly[styrene-*co*-(4-vinylbenzyl alcohol)] containing 10% hydroxyl group was used for grafting of ε-CL and acetylation with vinyl acetate. In these two reactions the steric environment of the benzyl group was distinguished. This type of reaction selectivity is impossible by a conventional chemical process. Those results can be extended to selective functionalization of polymers by lipase-catalyzed transesterification, that demands high stereochemical control [115].

Enantioselective Polymerization. An enantioselective ROP of 3-methyl-4-oxa-6-hexanolide (MOHEL) was catalyzed by lipase PC in bulk at 60°C (Scheme32) [86c]. The apparent initial reaction rate of *(S)*-isomer was seven times larger than that of *(R)*-isomer, indicating the effective enantioselective polymerization.

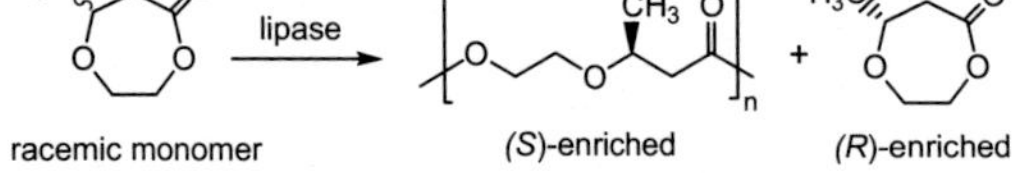

Scheme 32

Lipase PS-30 induced an enantioselective ROP of α-methyl-β-propiolactone (4-membered) in toluene to produce an optically active *(S)*-enriched (uo to 75% *(S)*) polymer with M_n values from 2,000 to 2,900 and $[\alpha]^{25}_D$ +12.2° to +19.0° (c 0.9 g/dL, CHCl$_3$) [116]. Enantioselective ROP of methyl substituted ε-CLs was carried out with Novozym 435 catalyst. A benzyl alcohol-initiated ROP at 45°C showed *(S)*-selective for 3-MeCL, 4-MeCL, and 6-MeCL and *(R)*-selective for 5-MeCL. 6-MeCL did not propagate in the ROP. Values of the enantiomeric ratio E of 3-MeCL, 4-MeCL, and 5-MeCL were 13 ± 4, 93 ± 27, and 27 ± 7, respectively. The enantiomeric excess (ee) value in the polymer of 4-MeCL was 0.88 [117].

In the lipase CA-catalyzed copolymerization of β-butyrolactone (β-BL) with DDL, *(S)*-β-BL was preferentially reacted to give the *(S)*-enriched optically active copolymer with ee of β-BL unit = 69%. δ-Caprolactone (6-membered) was also enantioselectively copolymerized with achiral lactones by the lipase catalyst to give the *(R)*-enriched optically active polyesters reaching 76% ee, in which the enantioselectivity was opposite [109b,118].

Lipase CA-catalyzed ROP of ω-methylated six lactones was studied for elucidation of the relationships between enantioselectivity and lactone structure (Scheme 33) [119]. The rate of polymerization was much affected by the ring size and the enantioselectivity was switched from *(S)*-selective for small (4-, 6-, and 7-membered) lactones to *(R)*-selective for large (8-, 9-, and 13-membered) lactones. From the k_{cat} values (s^{-1}) [for 4-, 6-, 7-, 8-, 9-, and 13-membered: *(S)*-enantiomers; 45.7, 7.9, 49.3, 0.01, nd, and nd, respectively; *(R)*-enantiomers; nd, 7.6, 8.5, 204.4, 10.3, and 23.3, respectively, where nd meaning 'not detected'], the enantiomeric ratio E was very large, in particular, for the large lactones. The selectivity switch was further supported by the molecular modeling studies on the free energy difference between the lactone structure and the active site cavity of the lipase. The lactone takes transoid and cisoid conformations; virtually small lactones for cisoid and large lactones for transoid. ROP of the small cisoid lactones was *(S)*-selective (3-MePL and 6-MeCL) or aselective (5-MeVL). ROP of the larger transiod lactones was *(R)*-selevtive with very high enantioselectivity. For the intermediate ring sizes, 7-MeHL and 8-MeOL, the significant amount of cisoid conformers present did not affect the enantioselectivity. The interpretation of the enantioselectivity and lactone structure relationship is attractive, but it seems actually not simple, particularly for understanding the intermediate ring sized case [119].

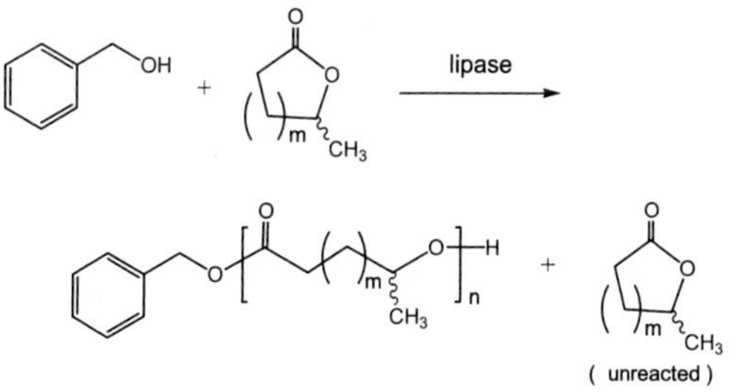

Scheme 33

A novel method of iterative tandem catalysis (ITC) was created, by which optically active oligoesters were obtained via ROP of 6-methyl-ε-CL (6-MeCL). ITC means a polymerization in which the chain growth is effectuated by a combination of two different catalytic processes that are both compatible and complementary. By combining the lipase CA-catalyzed ROP of racemic 6-MeCL and the Ru-catalyzed racemization of a propagating secondary alcohol in one-pot, optically pure oligoesters were obtained, which was developed based on the dynamic kinetic resolution (DKR) method (Scheme 34) [120]. first, lipase CA catalyzes the ring-opening of an *(S)* monomer enantioselectively to give the benzyl alcohol adduct, but the *(S)* alcohol is less favored to react with the monomer via ring-opening and hence the Ru-catalyzed racemization takes place to give a racemic alcohol. Then, an *(R)* alcohol selectively reacts with the monomer to facilitate one monomer-unit elongated. This reaction cycle repeats and ended up with the production of *(R)* oligoesters from the racemic monomer. The cycle could be repeated up to five times, being up to five monomer units of the product oligomers. Since the ring-opening of racemic 6-MeCL by lipase CA catalyst took place only for the *(S)*-isomer, *(R)*-6-MeCL monomer virtually remained unreacted.

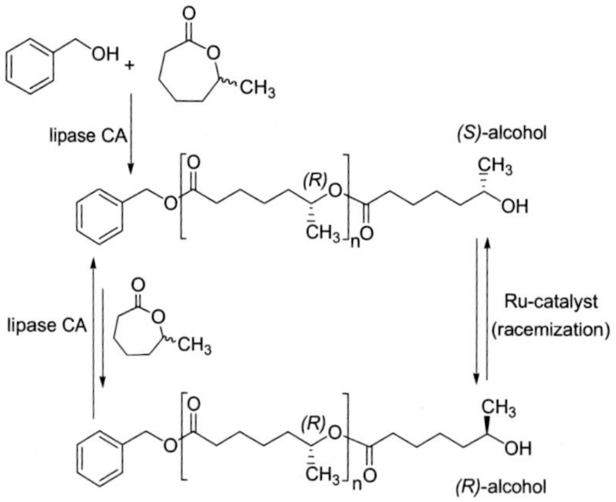

Scheme 34

Chemoenzymatic Polymerization. Enzyme is a "green" biocatalyst, in contrast to a "chemical" metal catalyst, which is sometimes toxic and not renewable. More importantly, catalysis of these two classes is quite different in function and mechanism, but can be mutually compatible each other to allow the catalysis concurrently in the same reaction system, if the combination of two classes of catalyst is appropriate. With utilizing these advantages of enzymes, chemoenzymatic method has been developed for the synthesis of new polymeric materials, in particular, various block copolymers, which are otherwise difficult to prepare.

By the combination of an enzymatic polymerization and a chemical polymerization a new chemoenzymatic method was developed; enzyme (lipase)-catalyzed ring-opening polymerization (eROP) of lactones and atom transfer radical polymerization (ATRP) (Scheme 35) [121]. The combination of two different, conceecutively-proceeding reactions is referred to as *cascade polymerization*. The method allows a versatile synthesis of block copolymers consisting from a polyester chain and a vinyl polymer chain by using a designed bifunctional initiator. For example, in route A an initiator having OH group for eROP and Br atom for ATRP is to be used. Lipase CA catalyzed the ROP of ε-CL at 60°C in toluene to give poly(ε-CL). Then, the poly(ε-CL) having the bromide was isolated and used for the Cu-catalyzed radical polymerization of styrene (St) at 85°C in 1,4-dioxane to give poly(ε-CL-*block*-St) in a high yield, in which the ε-CL chain exhibited M_n 5.8 × 10^3 and the St chain 15 × 10^3 [121]. During the reaction, methyl methacrylate interfered with eROP by transesterification to decrease the enzymetic activity, whereas *t*-butyl methacrylate did not [122].

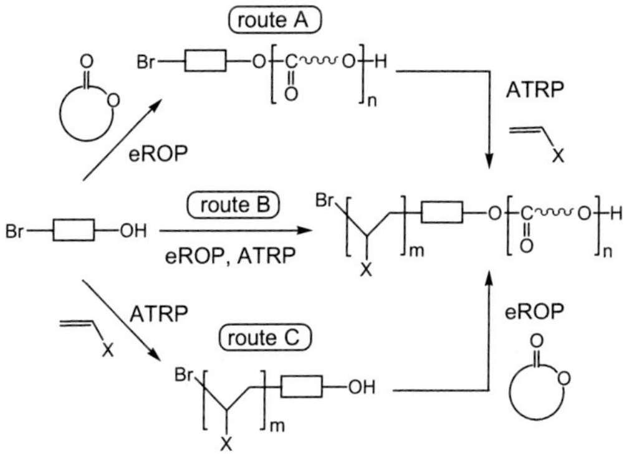

Scheme 35

A similar reaction was applied to an optically active block copolymer synthesis. For example, with using racemic monomer of 4-MeCL, lipase CA catalyst, and a bifunctional initiator having OH and Br, enantioselective enzymatic ROP of 4-MeCL was induced from the OH group and then ATRP of methyl methacrylate (MMA) was initiated from the Br atom to afford a block copolymer, poly(4-MeCL-*block*-MMA) having a chiral polyester chain [123]. Branched polymers with or without polyMMA chain from a poly(ε-CL) macromonomer were produced by using the chemoenzymatic technique: 2-Hydroxyethyl α-bromoisobutyrate was used as a bifunctional initiator to synthesize a Br containing poly(ε-CL) macromonomer, to which a polymerizable end group was introduced by in situ enzymatic acrylation with vinyl acrylate. Subsequent ATRP of the acrylate macromonomer gave branched polymers [124].

Similarly, the method was used for one-pot synthesis of a block copolymer by using a nitroxide mediated radical process. Thus, metal-free poly(ε-CL-*block*-St) was obtained in two consecutive polymerization steps via corresponding route B and in a one-pot cascade approach without intermediate transformation or work-up step via corresponding route A. A chiral block copolymer with high enantiomeric excess was obtained by using 4-alkyl ε-CL (4-RCL) and styrene as comonomers, where the first step was a lipase-catalyzed enantioselsctive ROP of 4-RCL at 60°C and the second was a free radical living polymerization of styrene at 95°C [125].

A spontaneous single-step chemoenzymatic synthesis of block copolymers was facilitated in supercritical carbon dioxide (scCO$_2$) via route A in Scheme 35 [126]. For example, the reaction was carried out in scCO$_2$; at 35°C, 1,500 psi (10.3 MPa) using MMA, ε-CL, Novozym 435, CuCl, and 2,2′-bipyridine. With reaction time of 20 h, a block copolymer, poly(ε-CL-*block*-MMA), of M_n = 41 000 (M_w/M_n = 2.11) was obtained in 60% yields. PolyMMA block part showed M_n = 10 000 (M_w/M_n = 1.02), indicating a living radical polymerization of MMA. Each homopolymer was produced in small amount, less than 10%. These data indicate that the two catalyst systems are robust under the reaction conditions and can tolerate each other. Block copolymer synthesis via routes B and C was also possible.

For chemoenzymatic synthesis combining eROP and ATRP, 2,2,2-trichloroetahnol was employed as a new initiator. first, CCl_3CH_2O-terminated poly(ε-CL) was prepared by Novozym 435-catalyzed ROP of ε-CL, and then, styrene (St) polymerization was initiated from C-Cl bond cleavage via ATRP to give rise to poly(ε-CL-*block*-St) having the degree of polymerization (DP) of ε-CL/St = 76/56. Instead of St, glycidyl methacrylate (GMA) was used for preparing a new amphiphilic block copolymer, poly(ε-CL-*block*-GMA), via chemoenzymatic technique. Products of two block copolymers possessed M_n of 22 800 with DP values ε-CL/GMA = 68/85 and M_n of 60 700 with DP values ε-CL/GMA = 68/352 [127].

A chemoenzymatic process combining eROP and ATRP was employed for synthesizing heterografted molecular bottle brushes (HMBB) having a complicated structure [128]. In the first step, an approximately 50% homografted polymer, poly(glycidol-*graft*-ε-CL-acetyl)-*co*-glycidol, was obtained via Novozym 435-catalyzed ROP of ε-CL using polyglycidol as a multifunctional initiator. Then, selective acetylation of the hydroxyl groups at the graft ends was performed via enzymatic acetylation with vinyl acetate, and the hydroxyl groups at the backbone were acylated with 2-bromo-2-methylpropionyl bromide. finally, MMA or *n*-BuMA grafts were attached by ATRP technique initiated from the bromine atom with CuBr/2,2′-bipyridyl catalyst at 60°C. The resulting HMBB possessed M_n values (reaction time): 5.5×10^4 (75 min) and 8.3×10^4 (120 min) for MMA grafts; 6.0×10^4 (75 min) and 8.3×10^4 (120 min) and 1.0×10^5 (210 min) for *n*-BuMA grafts.

For the synthesis of multifunctional poly(meth)acrylates, functional monomers of (meth)-acrylates were derived from Novozym 435-catalyzed transacylation of methyl methacrylate (MMA) and acrylate (MA) with various functional alcohols. These monomers thus prepared were radically polymerized by AIBN initiator to give poly(meth)acrylates via cascade reactions [129]. The obtained polymers possessed hydrophilic, hydrophobic, as well as cationic nature, and can be used for surface coatings in various ways.

ROP of Other Cyclic Monomers. Besides the lactone monomers, various cyclic monomers have been polymerized via lipase-catalyzed ROP (Figure 7). Only typical examples are shown here.

In 1993, a cyclic acid anhydride was found to undergo the ring-opening addition-condensation polymerization with a glycol by lipase PF catalyst, giving rise to a polyester having M_n ~2,000 (M_w/M_n = 1.4) in a good yield. During the reaction, the ring-opening as well as the dehydration occurred (Scheme 36(a)) [130]. Several cyclic anhydrides, succinic, glutaric, and diglycolic anhydrides, were polymerized by lipase CA catalyst with α,ω-alkylene glycols in toluene at 60°C to give the polyesters with M_n reaching 1.0×10^4. This ring-opening addition-condensation polymerization involving dehydration proceeded also in water and $scCO_2$ [131].

ROP of a new cyclic monomer of an *O*-carboxylic anhydride derived from lactic acid (lacOCA) has been achieved with catalysis of lipase (Scheme 36(b)) [132]. The polymerization proceeded within a few hours at 80°C in toluene with liberating carbon dioxide and gave poly(L-lactic acid) (PLLA) in high yields having high molecular weight, M_n up to 38 400 with low polydispersity $M_w/M_n < 1.4$. Slight preference in reactivity for L-lacOCA over D-lacOCA was observed.

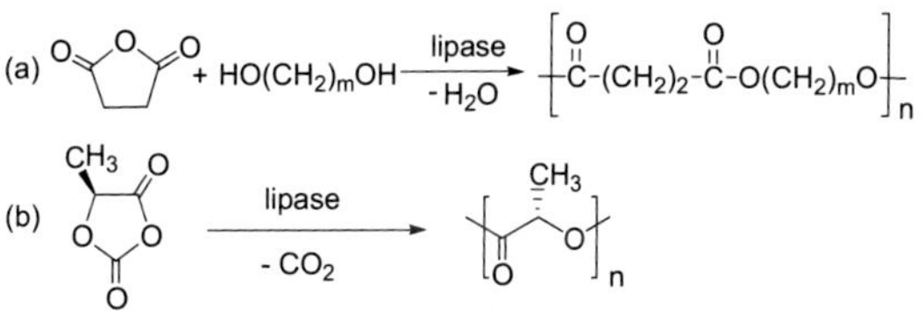

Scheme 36

In 1997, a polycarbonate from 1,3-dioxan-2-one (trimethylene carbonate, TMC) was first synthesized (Scheme 37(a)); lipase CA efficiently catalyzed ROP of TMC at 70°C to give the product with M_n higher than 1.0×10^4 [133]. At a higher temperature of 100°C, the ROP with PPL catalyst afforded a polycarbonate of higher M_w ~1.6×10^5. A new 7-membered cyclic carbonate monomer having a ketal group was derived from naturally occurring L-tartaric acid. Lipase-catalyzed ROP of the monomer was induced at 80°C in bulk, to afford the polycarbonate with M_n 15 500 with $M_w/M_n = 1.7$ (Scheme 37(b)). Deprotection of the ketal group resulted in optically pure polycarbonate (M_n 1.0×10^4, $M_w/M_n = 2.0$, $[\alpha]_D^{20} = +56°$) having hydroxy functional groups. The polycarbonate is considered to have potentials for various biomedical applications [134]. A degradable polycarbonate copolymer for pH-dependent controlled drug release with micelle formation was synthesized via a chemoenzymatic route. A triblock copolymer of ABA type was designed, where A block is poly(trimethylene carbonate) (PTMC) and B block is poly(PEG-co-cyclic acetal) (PECA), which was shown to form micelles and to be a biomaterial for drug carrier [135].

A 4-membered cyclic amide (β-lactam) was polymerized via Novozyme 435-catalyzed ROP to afford poly(β-alanine) with a degree of polymerization 8, having a linear structure (Scheme 37(c)) [136].

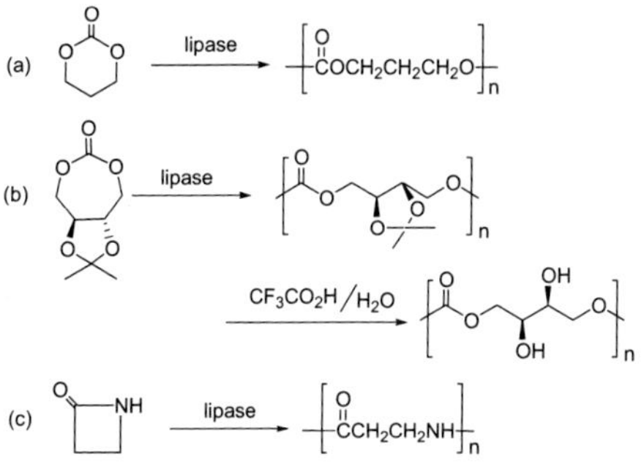

Scheme 37

A new 12-membered lactone, 2-oxo-12-crown-4-ether (OC), showed a high ROP reactivity by Novozym 435 catalyst to give polyOC (M_n 3,500) showing T_g ~ −40°C, soluble in water. Copolymerization of OC with PDL tuned the copolymer properties (M_n around 4,400–12 200). Although OC was polymerized five times faster than PDL, the copolymer showed a random copolymer structure [137].

4. 3. 5 Conclusions and future perspectives

In this mini, comprehensive rview, we have focused on enzymatic polymerization using hydrolases (hydrolysis enzymes) as catalysts. Therefore, syntheses of polysaccharides and polyesters are the main areas, and catalysts include mainly glycosidases and lipases. These hydrolysis enzymes catalyse the bond-cleavage with water *in vivo*, whereas in the polymerization they catalyze the bond-formation between monomers *in vitro*, a reverse direction of the *in vivo* reaction.

For the polysaccharide synthesis, the hydrolase catalysis enabled for the first time the *in vitro* synthesis of natural complicated biomacromolecules like cellulose, chitin, hyaluronan, chondroitin, and others. The synthesis of these compounds is very difficult via conventional methods.

Catalysis of the hydrolysis enzymes is highly selective in all respects, in enantio-, regio-, and chemo-selectivities, and can be achieved under mild reaction conditions. Lipase is active for selective end-functionalization (modification of polymers) with a clean process. The catalyst activity is achieved not only in naturally occuring solvent conditions, mainly in water, but also in pure organic media, in super critical carbon dioxide, and even in ionic liquids, at a wide range of temperature. Among others, it is a non-toxic biocatalyst, which is renewable. It is relatively cheap compared with other enzymes, and recently a variety of lipase became commercially available.

Because of these characteristics, lipase catalysis has potentials to contribute to environmental problems for maintaining green sustainable society by pursuing *green polymer chemistry* [2,3]. Some of the lipase-catalyzed synthesis and end-functionalization of polyesters mentioned above already provided with good examples of green polymer chemistry from the viewpoint of clean-process, energy savings, natural resources, carbon dioxide emission, *etc*.

For a possible way of conducting green polymer chemistry, a recycling of polyesters has been proposed [138]. Industrial examples of chemical recycling are few but known like an alcoholysis method of poly(ethylene terephthalate) (PET) and poly(butylene terephthalate). Also, a biobased polymer such as poly(lactic acid) (PLA) can be a good example in this direction. A method of chemical recycling of polymers using lipase catalysis was described [139]. The principle lies in that the ROP system of lactones by lipase catalysis is reversible between polymers and oligomers, which can be controlled by changing the reaction conditions, *i.e.*, the lactone polymerization gives cyclic oligomers including monomer in a dilute solution and gives higher molecular weight polymers in a concentrated solution which can be degraded to cyclic oligomers again in a dilute solution by the same catalyst.

References

1. P. T. Anastas, J. C. Warner, *"Green Chemistry: Theory and Practice"* Oxford University Press (1998)
2. (a) S. Kobayashi, *J. Polym. Sci. Part A: Polym. Chem.*, **37**, 3041–3056 (1999); (b) S. Kobayashi *et al.*, *Chem. Rev.*, **101**, 3793–3818 (2001); (c) S. Kobayashi, *Macromol. Rapid Commun.*, **30**, 137–166

(2009); (d) S. Kobayashi, A. Makino, *Chem. Rev.*, **109**, 5288–5353 (2009); (e) S. Kobayashi, *Proc. Jpn. Acad., Ser. B*, **86**, 338–365 (2010); (f) J. Kadokawa, S. Kobayashi, *Curr. Opin. Chem. Biol.*, **14**, 145–153 (2010); (g) A. Makino, S. Kobayashi, *J. Polym. Sci. Part A: Polym. Chem.*, **48**, 1251–1270 (2010)

3. J. E. Puskas *et al.*, *J. Polym. Sci. Part A: Polym. Chem.*, **47**, 2959–2976 (2009)
4. (a) J. B. Jones, *Tetrahedron*, **42**, 3351–3403 (1986); (b) C. H. Wong, G. M. P. Whitesides, *Enzymes in Synthetic Organic Chemistry*: Oxford (1994)
5. (a) S. Kobayashi *et al.*, In *Catalysis in Precision Polymerization*; S. Kobayashi, *ed.*; John Wiley & Sons: Chichester, England, pp 417–441 (1997); (b) S. Kobayashi *et al.*, *Bull. Chem. Soc. Jpn.*, **74**, 613–635 (2001); (c) R. A. Gross *et al.*, *Chem. Rev.*, **101**, 2097–2124 (2001); (d) S. Kobayash *et al.*, *Prog. Polym. Sci.* **26**, 1525–1560 (2001); (e) S. Kobayashi, H. Uyama, In *Encyclopedia of Polymer Science and Technology*; 3rd ed.; Kroschwitz, J. I., *ed.*; Wiley: New York, pp 328–364 (2003); (f) S. Kobayashi, *J. Polym. Sci. Part A: Polym. Chem.* **43**, 693–710 (2005). (g) S. Kobayashi, H. Ritter, D. Kaplan, *A special edition: Enzyme-Catalyzed Synthesis of Polymers, Adv. Polym. Sci.*, **194** (2006); (h) S. Kobayashi, M. Ohmae, In *Macromolecular Engineering: Precise Synthesis, Materials Properties, Applications*; Matyjaszewski, K., Gnanou, Y., Leibler, L., *eds.*; WILEY-VCH: Weinheim, Chapter 10; pp 400–477 (2007); (i) H. N. Cheng, R. A. Gross, *Polymer Biocatalysis and Biomaterials II, ACS Symposium Series* 999; Oxford University Press: Washington (2008); H. Ohara *et. al.*, *Biomacro molecules*, **12**, 3833–3837 (2011)
6. (a) L. Pauling, *Chem. Eng. News*, **24**, 1375 (1946); (b) P. A. Kollman *et al.*, *Acc. Chem. Res.* **34**, 72–79 (2001)
7. (a) S. Kobayashi *et al.*, *J. Am. Chem. Soc.*, **113**, 3079–3084 (1991); (b) S. Kobayashi, S. Shoda, *Int. J. Biol. Macromol.*, **17**, 373–379 (1995)
8. S. Kobayashi, *Proc. Jpn. Acad., Ser. B*, **83**, 215–247 (2007)
9. S. Egusa *et al.*, *Angew. Chem. Int. Ed.*, **46**, 2063–2065 (2007)
10. (a) I. Nakamura *et al.*, *Macromol. Biosci.*, **5**, 623–628 (2005); (b) I. Nakamura *et al.*, *Int. J. Biol. Macromol.*, **43**, 226–231 (2008)
11. F. Nakatsubo *et al.*, *J. Am. Chem. Soc.*, **118**, 1677–1681 (1996)
12. S. Kobayashi *et al.*, *Biomacromolecules*, **1**, 168–173 (2000)
13. (a) J. H. Lee *et al.*, *Proc. Natl. Acad. Sci., USA.*, **91**, 7425–7429 (1994); (b) S. Kobayashi *et al.*, *J. Macromol. Sci.; Pure Appl. Chem.*, **A34**, 2135–2142 (1997)
14. (a) T. Hashimoto *et al.*, *Biomacromolecules*, **7**, 2479–2482 (2006); (b) H.Tanak *et al.*, *Macromolecules*, **40**, 6304–6315 (2007)
15. E. Okamoto *et al.*, *Cellulose*, **4**, 161–172 (1997)
16. S. Kobayashi *et al.*, *Macromolecules*, **25**, 3237–3241 (1992)
17. B. Pfannemüller, *Int. J. Biol. Macromol.*, **9**, 105–108 (1987)
18. (a) J. Kadokawa *et al.*, *Chem. Commun.*, 449–450 (2001); (b) J. Kadokawa *et al.*, *Macromolecules*, **34**, 6536–6538 (2001); (c) Y. Kaneko *et al.*, *Macromolecules*, **41**, 5665–5670 (2008)
19. (a) Y. Kaneko *et al.*, *Biomacromolecules*, **8**, 2983–2985 (2007); (b) Y. Kaneko, J. I. Kadokawa, *J. Biomater. Sci.-Polym. Ed.*, **17**, 1269–1284 (2006); (c) Y. Kaneko, *et al.*, *Polym. J.*, **41**, 792–796 (2009)
20. S. Kobayashi *et al.*, *Macromolecules*, **29**, 2698–2700 (1996)
21. (a) S. Kobayashi *et al.*, *J. Am. Chem. Soc.*, **118**, 13113–13114 (1996); (b) T. Kiyosada *et al.*, *Polym. Prepr. Jpn.*, **44**, 1230–1231 (1995); (c) H. Sato *et al.*, *Anal. Chem.*, **70**, 7–12 (1998); (d) I. Tews *et al.*, *J. Am. Chem. Soc.*, **119**, 7954–7959 (1997)
22. (a) S. Shoda *et al.*, *Helv. Chimi. Acta*, **85**, 3919–3936 (2002); (b) M. Kohri *et al.*, *Holzforschung*, **60**, 485–491(2006)

23. J. Sakamoto, S. Kobayashi, *Chem. Lett.*, **33**, 698–699 (2004)
24. (a) H. Ochiai et al., *Carbohydr. Res.*, **339**, 2769–2788 (2004); (b) H. Ochiai et al., *Chem. Lett.*, **33**, 694–695 (2004)
25. (a) A. Makino et al., *Chem. Lett.*, **35**, 160–161 (2006); (b) A. Makino et al., *Macromol. Biosci.*, **6**, 862–872 (2006)
26. A. Makino et al., *Biomacromolecules*, **8**, 188–195 (2007)
27. A. Makino et al., *Polym. J.*, **38**, 1182–1188 (2006)
28. (a) S. Kobayashi et al., *J. Am. Chem. Soc.*, **123**, 11825–11826 (2001); (b) S. Kobayashi et al., *Macromol. Symp.*, **183**, 127–132 (2002)
29. (a) H. Ochiai et al., *Biomacromolecules*, **6**, 1068–1084 (2005); (b) H. Ochiai et al., *Biomacromolecules*, **8**, 1327–1332 (2007)
30. S. Kobayashi et al., *J. Am. Chem. Soc.*, **125**, 14357–14369 (2003)
31. S. Fujikawa et al., *Biomacromolecules*, **6**, 2935–2942 (2005)
32. M. Ohmae et al., *ChemBioChem*, **8**, 1710–1720 (2007)
33. M. Ohmae et al., *Macromol. Chem. Phys.*, **208**, 1447–1457 (2007)
34. M. Fujita et al., *J. Am. Chem. Soc.*, **120**, 6411–6412 (1998)
35. S. Kobayashi et al., *Biomacromolecules*, **7**, 1644–1656 (2006)
36. S. Kobayashi et al., *Macromol. Rapid Commun.*, **27**, 781–786 (2006)
37. A. Makino et al., *Biomacromolecules*, **7**, 950–957 (2006)
38. H. Ochiai et al., *Biomacromolecules*, **8**, 1802–1806 (2007)
39. A. Ajima et al., *Biotechnol. Lett.*, **7**, 303–306 (1985)
40. Y. Ohya et al., *J. Macromol. Sci.: Pure Appl. Chem.*, **A32**, 179–190 (1995)
41. S. Matsumura, J. Takahashi, *Makromol. Chem. Rapid Commun.*, **7**, 369–373 (1986)
42. D. O'hagan, N. A. Zaidi, *Polymer*, **35**, 3576–3578 (1994)
43. A. Olsson et al., *Biomacromolecules*, **8**, 757–760 (2007)
44. T. Ebata et al., *Macromol. Biosci.*, **8**, 38–45 (2008)
45. H. Ebata et al., *Macromol. Biosci.*, **7**, 798–803 (2007)
46. M. A. J. Veld et al., *J. Polym. Sci. Part A: Polym. Chem.*, **45**, 5968–5978 (2007)
47. D. Knani, D. H. Kohn, *J. Polym. Sci. Part A: Polym. Chem.*, **31**, 2887–2897 (1993)
48. U. Kanca et al., *J. Polym. Sci. Part A: Polym. Chem.*, **46**, 2721–2733 (2008)
49. S. D. Miao et al., *J. Polym. Sci. Part A: Polym. Chem.*, **46**, 4243–4248 (2008)
50. S. Okumura et al., *Agri. Biol. Chem.*, **48**, 2805–2808 (1984)
51. Z. L. Wang et al., *J. Macromol. Sci. Pure Appl. Chem.*, **A33**, 599–612 (1996)
52. A. Mahapatro et al., *Biomacromolecules*, **4**, 544–551 (2003)
53. H. Uyama et al., *Polym. J.*, **34**, 94–96 (2002)
54. (a) H. Uyama et al., *Chem. Lett.*, 1285–1286 (1998); (b) F. Binns et al., *J. Polym. Sci. Part A: Polym. Chem.*, **36**, 2069–2079 (1998)
55. B. Chen et al., *Biomacromolecules*, **9**, 463–471 (2008)
56. (a) S. Kobayashi et al., *Chem. Lett.*, 105 (1997); (b) S. Suda et al., *Proc. Jpn. Acad. B*, **75**, 201–206 (1999)
57. (a) A. Kumar et al., *Macromolecules*, **36**, 8219–8221 (2003); (b) A. S. Kulshrestha et al., *Macromolecules*, **38**, 3193–3204 (2005)
58. A. S. Kulshrestha et al., *Biomacromolecules*, **8**, 1794–1801 (2007)
59. M. Hunsen et al., *Macromolecules*, **40**, 148–150 (2007)
60. H. Uyama et al., *Chem. Lett.*, 893–894 (1999)
61. J. S. Wallace, C. J. Morrow, *J. Polym. Sci. Part A: Polym. Chem.*, **27**, 2553–2567 (1989)
62. C. Berkane et al., *Macromolecules*, **30**, 7729–7734 (1997)

63. H. Azim et al., *Biomacromolecules*, **7**, 3093–3097 (2006)
64. R. Marcilla et al., *Eur. Polym. J.*, **42**, 1215–1221 (2006)
65. I. Hilker et al., *Angew. Chem.-Int. Ed.*, **45**, 2130–2132 (2006)
66. H. Uyama, S. Kobayashi, *Chem. Lett.*, 1687–1690 (1994)
67. U. Uyama et al., *J. Polym. Sci. Part A: Polym. Chem.*, **37**, 2737–2745 (1999)
68. T. Takamoto et al., *e-Polymers*, **4**, 1–6 (2001)
69. H. Uyama et al., *Polym. J.*, **31**, 380–383 (1999)
70. B. J. Kline et al., *J. Am. Chem. Soc.*, **120**, 9475–9480 (1998)
71. H. Uyama et al., *Macromol. Biosci.*, **1**, 40–44 (2001)
72. H. Uyama et al., *Chem. Lett.*, 800–801 (2000)
73. S. Namekawa et al., *Biomacromolecules*, **1**, 335–338 (2000)
74. Z. Z. Jiang, *Biomacromolecules*, **9**, 3246–3251 (2008)
75. T. Takamoto et al., *Macromol. Biosci.*, **1**, 223–227 (2001)
76. (a) T. Tsujimoto et al., *Biomacromolecules*, **2**, 29–31 (2001); (b) T. Tsujimoto et al., *Macromol. Biosci.*, **2**, 329–335 (2002)
77. H. Uyama et al., *Biomacromolecules*, **4**, 211–215 (2003)
78. M. Kato et al., *Biomacromolecules*, **10**, 366–373 (2009)
79. D. Knani et al., *J. Polym. Sci. Part A: Polym. Chem.*, **31**, 1221–1232 (1993)
80. (a) H. Uyama, S. Kobayashi, *Chem. Lett.*, 1149–1150 (1993); (b) H. Uyama et al., *Proc. Jpn. Acad. B*, **69**, 203–207 (1993)
81. A. C. Albertsson, I. K. Varma, *Biomacromolecules*, **4**, 1466–1486 (2003)
82. (a) S. Matsumura et al., *Chem. Lett.*, 795–796 (1996); (b) S. Namekawa et al., *Polym. J.*, **28**, 730–731 (1996); (c) Y. Y. Svirkin et al., *Macromolecules*, **29**, 4591–4597 (1996); (d) G. A. R. Nobes et al., *Macromolecules*, **29**, 4829–4833 (1996)
83. A. A. Panova et al, *Biomacromolecules*, **4**, 19–27 (2003)
84. S. Matsumura et al., *Macromol. Rapid Commun.*, **18**, 477–482 (1997)
85. M. Hans et al., *Macromol. Biosci.*, **9**, 239–247 (2009)
86. (a) H. Uyama et al., *Acta Polym.*, **47**, 357–360 (1996); (b) H. Uyama et al., *Polym. J.*, **29**, 299–301 (1997); (c) S. Kobayashi et al., *Polym. Degrad. Stabil.*, **59**, 195–201 (1998)
87. (a) R. T. MacDonald et al., *Macromolecules*, **28**, 73–78 (1995); (b) L. A. Henderson et al., *Macromolecules*, **29**, 7759–7766 (1996); (c) S. Matsumura et al., *Macromol. Rapid Commun.*, **21**, 860–863 (2000); (d) H. Ebata et al., *Biomacromolecules*, **1**, 511–514 (2000); (e) Y. Mei et al., *Macromolecules*, **35**, 5444–5448 (2002); (f) A. Cordova et al., *Polymer*, **40**, 6709–6721 (1999); (g) S. Divakar, *J. Macromol. Sci.: Pure Appl. Chem.*, **A41**, 537–546 (2004); (h) A. Kumar, R. A. Gross, *Biomacromolecules*, **1**, 133–138 (2000); (i) S. Kobayashi et al., *Macromol. Chem. Phys.*, **199**, 1729–1736 (1998); (j) H. Kikuchi et al., *Polym. J.*, **34**, 835–840 (2002); (k) K. Küllmer et al., *Macromol. Rapid Commun.*, **19**, 127–130 (1998)
88. (a) H. Uyama et al., *Chem. Lett.*, 1109–1110 (1997); (b) G. Sivalingam, G. Madras, *Biomacromolecules*, **5**, 603–609 (2004)
89. M. Hans et al., *Macromolecules*, **39**, 3184–3193 (2006)
90. Y. Soeda et al, *Macromol. Biosci.*, **5**, 277–288 (2005)
91. S. Kobayashi et al., *Macromolecules*, **31**, 5655–5659 (1998)
92. (a) H. Uyama et al., *Bull. Chem. Soc. Jpn.*, **68**, 56–61 (1995); (b) H. Uyama et al., *Macromolecules*, **28**, 7046–7050 (1995); (c) K. S. Bisht et al., *Macromolecules*, **30**, 2705–2711 (1997); (d) S. Namekawa et al., *Proc. Jpn. Acad. B*, **74**, 65–68 (1998)
93. Z. Z. Jiang et al., *Biomacromolecules*, **8**, 2262–2269 (2007)
94. W. Gao et al., *Macromolecules*, **40**, 145–147 (2007)

95. S. Namekawa *et al.*, *Polym. J.*, **30**, 269–271 (1998)
96. A. Taden *et al.*, *Macromol. Rapid Commun.*, **24**, 512–516 (2003)
97. (a) F. C. Loeker *et al.*, *Macromolecules*, **37**, 2450–2453 (2004); (b) C. J. Duxbury *et al.*, *J. Am. Chem. Soc.*, **127**, 2384–2385 (2005)
98. H. Uyama *et al.*, *Acta Polym.*, **49**, 700–703 (1998)
99. R. K. Srivastava, A. C. Albertsson, *Macromolecules*, **39**, 46–54 (2006)
100. B. Kalra *et al.*, *Macromolecules*, **37**, 1243–1250 (2004)
101. (a) K. S. Bisht *et al.*, *J. Am. Chem. Soc.*, **120**, 1363–1367 (1998); (b) A. Cordova *et al*, *J. Am. Chem. Soc.*, **120**, 13521–13522 (1998)
102. M. de Geus *et al.*, *Biomacromolecules*, **9**, 752–757 (2008)
103. K. R. Yoon *et al.*, *Adv. Mater.*, **15**, 2063–2067 (2003)
104. H. Uyama *et al.*, *Chem. Lett.*, 1047–1048 (1995)
105. H. Uyama *et al.*, *Bull. Chem. Soc. Jpn.*, **70**, 1691–1695 (1997)
106. M. Takwa *et al.*, *Macromol. Rapid Commun.*, **27**, 1932–1936 (2006)
107. N. Simpson *et al.*, *Macromolecules*, **41**, 3613–3619 (2008)
108. A. Duda *et al.*, *Macromolecules*, **35**, 4266–4270 (2002)
109. (a) S. Namekawa *et al.*, *Int. J. Biol. Macromol.*, **25**, 145–151 (1999); (b) S. Kobayashi, *Macromol. Symp.*, **240**, 178–185 (2006)
110. L. van der Mee *et al.*, *Macromolecules*, **39**, 5021–5027 (2006)
111. (a) H. Uyama *et al.*, *Macromolecules*, **34**, 6554–6556 (2001); (b) S. Habaue *et al.*, *Polymer*, **44**, 5195–5200 (2003)
112. I. van der Meulen *et al.*, *Biomacromolecules*, **9**, 3404–3410 (2008)
113. R. Kumar, R. A. Gross, *J. Am. Chem. Soc.*, **124**, 1850–1851 (2002)
114. M. Fujioka *et al.*, *Macromol. Rapid Commun.*, **25**, 1776–1780 (2004)
115. M. Padovani *et al.*, *Macromolecules*, **41**, 2439–2444 (2008)
116. Y. Y. Svirkin *et al.*, *Macromolecules*, **29**, 4591–4597 (1996)
117. J. Peeters *et al.*, *Biomacromolecules*, **5**, 1862–1868 (2004)
118. H. Kikuchi *et al.*, *Macromolecules*, **33**, 8971–8975 (2000)
119. J. van Buijtenen *et al.*, *J. Am. Chem. Soc.*, **129**, 7393–7398 (2007)
120. (a) B. A. C. van As *et al.*, *J. Am. Chem. Soc.*, **127**, 9964–9965 (2005); (b) J. van Buijtenen *et al.*, *Chem. Commun.*, 3169–3171 (2006)
121. U. Meyer *et al.*, *Macromolecules*, **35**, 2873–2875 (2002)
122. M. de Geus *et al.*, *J. Polym. Sci. Part A: Polym. Chem.*, **44**, 4290–4297 (2006)
123. J. Peeters *et al.*, *Biomacromolecules*, **5**, 1862–1868 (2004)
124. J. W. Peeters *et al.*, *Macromol. Rapid Commun.*, **26**, 684–689 (2005)
125. B. A. C. van As *et al.*, *Macromolecules*, **37**, 8973–8977 (2004)
126. C. J. Duxbury *et al.*, *J. Am. Chem. Soc.*, **127**, 2384–2385 (2005)
127. (a) K. Sha *et al.*, *J. Polym. Sci. Part A: Polym. Chem.*, **45**, 5037–5049 (2007); (b) K. Sha *et al.*, *J. Polym. Sci. Part A: Polym. Chem.*, **44**, 3393–3399 (2006)
128. M. Hans *et al.*, *Macromolecules*, **40**, 8872–8880 (2007)
129. D. Popescu *et al.*, *Macromol. Chem. Phys*, **210**, 123–139 (2009)
130. S. Kobayashi, H. Uyama, *Makromol. Chem. Rapid Commun.*, **14**, 841–844 (1993)
131. H. Uyama *et al.*, *Biochem. Eng. J.*, **16**, 145–152 (2003)
132. C. Bonduelle *et al.*, *Biomacromolecules*, **10**, 3069–3073 (2009)
133. (a) S. Kobayashi *et al.*, *Macromol. Rapid Commun.*, **18**, 575–579 (1997); (b) S. Matsumura *et al.*, *Macromolecules*, **30**, 3122–3124 (1997); (c) K. S. Bisht *et al.*, *Macromolecules*, **30**, 7735–7742 (1997)
134. R. Wu *et al.*, *Biomacromolecules*, **9**, 2921–2928 (2008)

135. S. Kaihara *et al.*, *Macromol. Biosci.*, **9**, 613–621 (2009)
136. L. W. Schwab *et al.*, *Macromol. Rapid Commun.*, **29**, 794–797 (2008)
137. L. van der Mee *et al.*, *J. Polym. Sci. Part A: Polym. Chem.*, **44**, 2166–2176 (2006)
138. S. Kobayashi *et al.*, *Biomacromolecules*, **1**, 3–5 (2000)
139. (a) S. Matsumura, *Macromol. Biosci.*, **2**, 105–126 (2002); (b) Y. Takahashi *et al.*, *Macromol. Biosci.*, **4**, 346–353 (2004); (c) A. Kondo *et al.*, *Macromol. Biosci.*, **8**, 533–539 (2008)

Chapter 5
Application of Bio-based Polymers

Masatsugu Mochizuki

5. 1   Introduction

For the past 80 years, synthetic polymers such as nylon, polyester, and polypropylene made from petroleum feedstock have been used for industrial applications as well as for clothing. However, the high economic growth of the 20 century has led big problems, global worming in addition to exhaustion of fossil resources and solid waste management of the petroleum-based polymers. In recent years bio-based polymers from plants have been focused upon attention as carbon-neutral materials without increase in greenhouse gas emission in solid waste management, for example in incineration.

There are many types of bio-based polymers including thermoplastics, thermosetting plastics, and non-thermoplastics such as regenerated cellulose or chitin. Here I will restrict myself to application of bio-based thermoplastics which normally can be applied for extensive molding processes such as extrusion, injection, thermoforming, foaming, and blow molding.

Polylactic acid (PLA) has been highlighted as a carbon-neutral bioplastic because of its availability from agricultural renewable resources like corn. With PLA, CO_2 is removed from the atmosphere when growing the feedstock crop, and returned to the Earth when PLA is degraded. Since the process recycles the Earth's carbon, PLA has potential to suppress atmospheric CO_2 levels. Also PLA is a biodegradable aliphatic polyester with thermoplastic processability, and has favorable both mechanical properties and biodegradation characteristics [1–3]. Cargill Dow LLC (now NatureWorks™ LLC) started full commercial manufacture of NatureWorks™ PLA with reducing production costs in 2002.

PLA is the most potential bio-based polymers as a commodity plastic having excellent balance of properties and environmental benefits, although PLA has enjoyed little success in replacing petroleum-based plastics outside of biomedical applications such as a bolt for bone connection until recently. Technological improvements to modify its heat resistance, hydrolytic stability and impact strength to widen its material performance have been underway [4,5]. Now PLA products are finding uses in many applications, including packaging films/sheets, fibers/nonwovens and a host of molded articles.

5.2 Key performance features of PLA

5.2.1 Chemical, physical and thermal properties

PLA is a crystalline thermoplastic polymer of aliphatic polyester, therefore it contains no aromatic ring structure. However PLA is of relatively high modulus with high gloss and clarity. PLA has a relatively high tensile modulus of 3500–4000 MPa, this being comparable with that of cellophane and oriented poly (ethylene terephthalate) (PET). Currently commercially available PLA is comparatively thermally stable, as indicated by a glass transition temperature (T_g) of 57–63°C, a crystallization temperature (T_c) of 110°C, and melting temperature (T_m) of 130–190°C which depends on the content of D-isomer in the PLA polymer. The physical properties of PLA depend primarily on their molecular characteristics, highly ordered structures and morphology.

PLA has a lower density (1.25 g/m^3) than that (1.34 g/m^3) of PET. PLA is inherently a moderately polar material due to the basic repeat units of ester bonds. This leads to a number of unique attributes such as resistance to both water and oils, similar to PET. PLA is basically hydrophobic, but more hydrophilic than PET. The moisture regain of PLA is 0.5 weight per cent, higher than the 0.3 weight per cent of PET. PLA is liable to hydrolysis, however, under high temperature/humid conditions, over 50°C /85 per cent RH.

5.2.2 Biodegradability and their biodegradation mechanism

Under typical use and storage conditions such as at room temperature, PLA is stable. However, under very specific conditions of high temperature (above 60°C) and high humidity (above 80 per cent RH), typified by the conditions for composting, PLA will disintegrate within one week to one month, followed by bacterial attack on the fragmented residues to give carbon dioxide and water. PLA is completely biodegradable when exposed in biologically active environments such as municipal compost facilities, along with other compostable organic materials. A typical degradation curve of PLA under composting conditions is shown in Figure 1 [6].

In the primary degradation phase, PLA undergoes chemical hydrolysis, which is both temperature- and humidity-dependent and does not involve any microorganisms. As the M_n reaches approximately 10 000–20 000, microorganisms present in the compost begin to digest the lower molecular weight oligomer and lactic acid, producing carbon dioxide and water. This two-stage biodegradation mechanism differs distinctly from that of many other biodegradable polymers such as poly(3-hydroxyalkanoate)s (PHA) and poly(butylene succinate) (PBS) which degrade by a single-step surface erosion process, involving direct bacterial attack with enzymatic degradation on the polymer [7,8]. Few reports exist on the microbial degradation of high molecular weight of PLA.

The unique two-stages/two-types of degradation mechanism of PLA suggests that it has a potential for not only bio-recycling but also structural materials with long term durability, because the 1st stage of chemical hydrolysis is rate-determined. That is, no degradation of chemical hydrolysis occurs under ordinary conditions around room temperature for practical uses as a structural material, but both chemical hydrolysis and biodegradation occur under high temperature/humid conditions such as composting as a bio-recycling material [5].

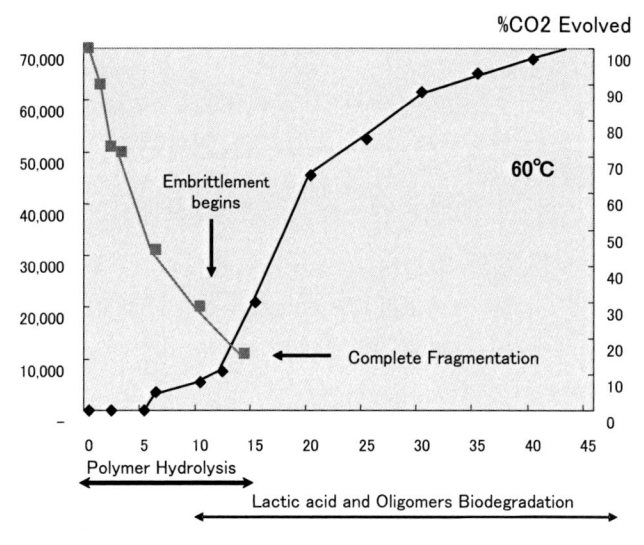

Figure 1 Biodegradation behavior of PLA at 60°C in compost [6]

5. 2. 3 Environmental sustainability

Conventional synthetic polymers rely on oil reserves for their feedstock source, the main drawbacks being not only that fossil resources will be exhausted after *ca.* 50 years but also that their use has the harmful impact on the environment with an increase in greenhouse gas. In contrast, the monomer in PLA is derived from renewable resources such as agricultural crops that are grown and harvested annually. With PLA, CO_2 is removed from the atmosphere when growing the feedstock crop, and returned to the Earth when PLA is degraded or incinerated.

Since the process recycles the Earth's carbon, PLA has the potential for not increasing atmospheric CO_2 levels. However, as is the case for all polymers, fossil fuels are used in the processing of raw materials and resin production for PLA. Thus, the subtotal CO_2 emission (feedstock + processing) of PLA pellet production has been evaluated to be the lowest, compared with that of conventional petroleum-based plastics [9]. Furthermore, the CO_2 emission of PLA in the course of disposal, such as incineration or composting, is minimal, so that the total CO_2 emission of PLA from cradle to grave (feedstock + processing + disposal) is the lowest among existing plastics [10].

5. 3 Processing of PLA

5. 3. 1 Melt crystallization and cold crystallization

Processing of thermoplastics is primarily solidification and melt crystallization during cooling process from melting point to room temperature. In particular, moldings of crystalline polymers are closely involved in two types of crystallization, melt crystallization or cold crystallization, as shown in Table 1.

Injection moldings are, however, the most difficult process to crystallize because both nucleation and growth of crystals must be induced during cooling process at the rapid rate of

Table 1 Crystallization in the course of thermoplastic processing

Melt Crystallization	Cold Crystallization
Crystallization during cooling from the melt to crystallization temperature	*Crystallization during heating from R. T. to crystallization temperature*
Injection molding	Thermoforming
Extrusion molding (fibers & films) → *followed by* Drawing & Annealing	

cooling, ranging 10^4–10^5°C/min. PLA can't be normally crystallized in injection moldings due to the slower rate of melt crystallization. This is the reason why injection molded articles have poor heat resistance as low as 55°C.

On the other hand, thermoforming is cold crystallization process in which polymer can be crystallized during heating from room temperature to crystallization temperature, once after the extrusion of sheet. However, a conventional PLA can hardly crystallize as well during thermoforming, although the cold crystallization is not so difficult as compared with melt crystallization.

In case of fibers/films making, subsequent drawing and annealing process after the extrusion of them will readily promote flow-induced crystallization of the highly oriented polymer chains, because nuclei or precursors of crystals had been formed during the first extrusion and cooling process of them.

In accordance with the classical theory of polymer crystallization, the spherulite growth rate G is given by

$$G = G_0 \exp(-E_D/RT - KT_m/R(T_m-T)) \tag{1}$$

where G_0 is constant, E_D is the activation energy for diffusion of polymeric segments, K is the energy factor for nucleation, R is the gas constant, T is crystallization temperature, and T_m is the melting point [11]. The first term of equation (1) is correlated to the diffusion process of polymeric segments to crystallize, and generally it has a positive correlation between G and T. The second term is correlated to the nucleation process, and it is generally inversely proportional to T. Consequently, we have the optimum crystallization temperature T_c between T_g and T_m, in which the rate of crystallization is maximum.

5. 3. 2 Crystallization rate of PLA

Currently commercially available PLA is not a homopolymer of L-lactic acid but a random copolymer composed of large amounts of L-lactic acid and small amounts of D-lactic acid. Of a particular interest here, is the dependence of the D-isomer units concentration, the mole fraction of D-isomer units, X_D, in the random copolymer on the crystallization behavior. The interest lies within the ability of the PLA polymers to crystallize more rapidly and form highly ordered structure with higher melting temperature, in other words the effects of stereo-regularity of the PLA polymeric chains as the structural factors on the crystallization behavior.

Recently we found that there was a monotonic increase in both the rate of crystallization and crystallinity of the PLA random copolymers of similar weight-average molecular weight,

M_w, with decreasing X_D in the differential scanning calorimetry measurement, as shown in Table 2 [12].

The relationship between half-crystallization time, $t_{1/2}$, and X_D, indicates that a half in $t_{1/2}$ for a 1% decrease in X_D. In addition, about 20°C increase in T_m is expected for a 1% decrease in X_D for currently commercially available PLA with D-isomer content of 1.4%.

We also have studied on the relationship between $t_{1/2}$ and crystallization temperature for PLA with different D-isomer contents, L-1, L-2, and L-3 respectively. The crystallization temperature at which crystallization rate is the maximum, in other word $t_{1/2}$ is the minimum, around 105–110°C as shown in Figure 2, while the crystallization rate increases with decreasing D-isomer contents of PLA.

Table 2 Crystallization parameters of PLA for isothermal melt crystallization at 110°C

Code	M_w ($\times 10^5$)	X_D (%)	t_s (min)	$t_{1/2}$ (min)	t_e (min)	t_e-t_s (min)	ΔH (J/g)	W_c (%)	k	n
L-1	1.39	0.25	0.971	3.02	8.12	7.15	35.15	37.8	2.37×10^{-2}	3.01
L-2	1.55	1.18	2.474	8.04	16.48	14.01	27.45	31.9	7.26×10^{-4}	3.28
L-3	1.42	2.21	5.192	14.23	28.69	23.50	22.02	23.7	3.50×10^{-5}	3.54

M_w: weight-average molecular weight, X_D: D-isomer contents of PLA
t_s: starting time of crystallization, $t_{1/2}$: half-crystallization time
t_e: ending time of crystallization, ΔH: enthalpy of crystallization, W_c: crystallinity
k: rate constant of crystallization in Avrami's equation., n: Avrami's Exponent

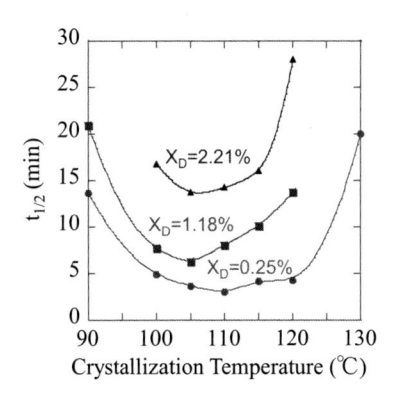

Figure 2 Relationship between half-crystallization time, $t_{1/2}$ and crystallization temperature of PLA with different D-isomer contents, X_D

5.4 High-performance PLA

We have two technical approaches for high-performance PLA. One is improvements of PLA itself without any oil-based engineering plastics. Another one is polymer alloy of PLA with oil-based engineering plastics like polycarbonate [5]. It is needless to say that the former is better than the latter from the view point of the reduction of carbon dioxide emission. However, we can't keep out of the latter way in order to design flame-retardant PLA or high impact strength PLA. The biomass carbon content of these materials is more than 30% that is coming up to the standard of more than 25% for 'BiomassPla' symbol mark by JBPA (Japan BioPlastics Association).

5. 4. 1 Heat resistance

It has been known that the rate of crystallization of PLA in injection moldings is too slow to give a contribution to heat resistance for practical uses. In 2002, a successful method for enhancement of heat resistance of PLA has first been developed by Unitika Ltd by a novel technology of PLA/clay nanocomposite, in which a hundred-fold increase in the rate of crystallization of PLA was dramatically achieved [4,5]. In isothermal crystallization of PLA at 130°C, the PLA/clay nanocomposite could start to crystallize within 1 minute, while a conventional PLA neat resin could at last crystallize after 100 minutes. It was found that DTUL (Distortion Temperature Under Load) under 0.45 MPa of the PLA/clay nanocomposites molded is 120–150°C, that was comparative to or better than that of oil-based commodity plastics such as polystyrene, polypropylene and ABS.

5. 4. 2 Hydrolysis resistance

PLA is also known to be chemically hydrolyzed slowly even at room temperature, and PLA products have normally 3–5 years of shelf life. In addition to heat resistance, long term durability from five to ten years under ordinary condition has been required for practical uses of returnable tablewares, housing of electronics devices, and automotive interior parts. The improvement in durability was achieved by enhancing hydrolysis resistance of PLA under high temperature above 55°C and high-humid above 95% RH conditions.

In respect of the techniques for enhancing hydrolysis resistance, it is first essential for us to suppress residual lactides in less than 2,000 ppm in PLA to block the chemical hydrolysis at room temperature. Then, one of the most effective approaches is capping of free COOH end groups of PLA, which is supposed to catalyze and accelerate the chemical hydrolysis, by adding oxazole or carbodiimide compounds to PLA. As a result, it was found that the PLA compound resin had retained a 90% of the initial strength after 1,000 hrs exposure under 60°C/95% RH condition, whereas a conventional PLA was steeply depressed to a zero % of the initial strength even after 200 hrs [4,5].

5. 4. 3 Impact strength

It is well known that PLA is rigid and brittle, and is therefore short of toughness and impact strength which are a key performance of for instance housing of mobile phone. The most familiar way to raise flexibility and toughness of PLA is to incorporate plasticizers or impact modifiers such as glycerol esters, adipic acid esters, lactic acid esters, rosin derivatives, or AB type of block copolymer composed of aliphatic polyester and PLA. A four-fold increase in impact strength, 9.6 KJ/m^2 for Izod, was observed for the improved PLA compound resin for injection molding by blending plasticizers and fibrous fillers as a reinforcement [5].

5.5 PLA products and potential applications

In this article, the key performance features and potential applications of PLA products including fibers/nonwovens, films/sheets, injection, thermoforming, foaming, and blow moldings, which are currently commercially available under the trademark TERRAMAC™ of Unitika Ltd, Japan, will be described [5,13].

5.5.1 Fibers and nonwovens

As before-mentioned, PLA may be crystallized relatively quickly by drawing or stretching with annealing due to molecular orientation followed by crystallization, yielding enhanced physical and thermal properties. Thus highly oriented fibers with good mechanical properties can be produced by melt spinning followed by subsequent drawing operations (two-step method), continuous spinning/drawing or high-speed spinning (one-step method). An increased tensile strength and modulus accompany crystallization development, and give excellent yarn properties such as tenacity, toughness, dimensional stability, and heat resistance, which are a little bit lower than those of PET.

PLA is an aliphatic polyester having no aromatic ring structure, but the yarn properties of PLA are relatively similar to those of PET. However, the density and refractive index of PLA is lower than those of PET, these being light-weight and deep silky-luster without brightness. The moisture regaining and wicking properties of PLA are superior to those of PET because PLA is more hydrophilic than PET. In addition the lower modulus leads to better properties of drape and hand-feel, whilst the crimp retention property leads to excellent crease resistance [2].

PLA fibers are not only biodegradable but also highly functional ones, coupled with their intrinsic properties such as bacteriostatic, flame-retardant, and weather-resistant properties, when compared with conventional polyester (PET) fibers [13,14].

In natural environments such as in soil or in water, the degradation proceeds slowly but steadily, this being a convenient feature for the application of agricultural/horticultural and geotextiles. After 2 years, the fibers have lost about 50% of their initial strength. In compost at 58°C, however, PLA fibers have degraded more rapidly in ten days than that natural fibers like wool or cotton have degraded.

We have found first that PLA fibers have bacteriostatic property under the standard test method using *Staphylococcus aureus* ATCC 6538P as inoculums. This unique advantage may be resulted from antimicrobial property of lactic acid resided in extremely small quantities in polymer, and offers benefits in the application fields of food, sanitation and agriculture.

The weather resistance of PLA fibers was evaluated using accelerated weathering test (Sunshine Weather Meter) in comparison with PET fibers. The weather resistance of PLA was superior to PET, suggesting PLA can be significant players as agricultural and geotexitiles in the natural environments where the weather resistance is required.

We found that PLA fibers and spunbond fabrics had higher LOI (Limit of Oxygen Index; JIS K7201) value ranging 23–30 than 20–21 of PET fibers, showing an excellent flame-retardant and self-extinguishing property with lower smoke generation. PLA conjugated (side by side) hollow fibers (HP8F) for fiberfill without any non-flammable chemicals have been approved as the flame-retardant product by Flame-retardant Products Association, JAPAN.

Potential applications of these fibers and nonwovens include geotextiles (vertical drain sheets, sand bags, erosion protection and slope stabilization), agricultural/horticultural products (plant covers, plant pots, strings), industrial uses (textiles & nonwovens, filters, cabin parts of vehicles), home furnishings & clothing (towels, bedding, furniture wadding and filling, carpets, T-shirts, casual wear, filters for teabags or organic wastes), and hygienic products (wipes, disposable diapers, personal care products) *etc*.

5. 5. 2 Films and sheets

In respect of PLA packaging applications, there are two specific areas, namely high-value films and rigid thermoformed containers. The first one is biaxially-oriented PLA films which offers good toughness and stiffness as well as excellent gloss and transparency, similar to oriented PET films and cellophane. PLA with D-isomer level below 8% can be semi-crystalline if nucleated, annealed, or subjected to strain-induced orientation processes. Enhanced toughness and heat resistance accompanies crystallinity development. The melting point of PLA increases with decreasing content of the D-isomer, and ranges from about 130–190°C [12,13].

Oriented PLA film is shown to have excellent shrinkage property for packaging and labeling. Furthermore, characteristics of excellent deadfold and twist retention are comparable to with those of cellophane films. PLA is an inherently polar material due to the repeat unit of lactic acid, leading to the benefit of resistance to grease or oil. PLA has a high critical surface energy, making it receptive to holding corona treatment well, and therefore easily printable and metallizable [2].

The second one of PLA packaging applications is unoriented sheet as extruded which has extremely lower haze and good moldability in thermoforming, bringing promise to the area of food containers and blisters for replacement of amorphous PET (A-PET).

PLA films and sheets are mainly applied for packaging materials (films for wrapping, food dishes and trays, cups, fruits & vegetables containers, blisters, paper lamination). Compost bag for organic wastes and agricultural mulch film are also potential application of PLA films.

5. 5. 3 Injection molding

PLA can't be normally crystallized in ordinary injection molding process due to the slower rate of melt crystallization, leading to the molded articles with poor heat resistance as low as 55°C. The novel technology of PLA/clay nanocomposite with exceedingly higher rate of crystallization in mold at 110°C provided a real breakthrough towards PLA products with high heat resistance above 120°C.

Recently, high-performance PLA as durable structural materials with heat resistance and hydrolysis resistance have been developing, and are actually employed in injection molding as a returnable tableware (Figure 3), a housing of electronics device like mobile phones (NTT DoKoMo/NEC, FOMA N701i ECO, Figure 4), a part of business machine, automotive interior, and miscellaneous goods for life.

Another one is polymer alloy of PLA with oil-based plastics such as polycarbonate or ABS

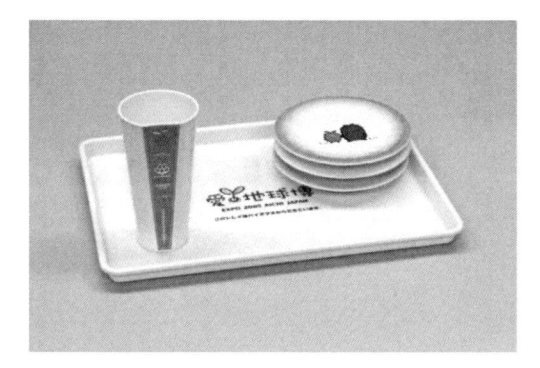

Figure 3 Returnable tableware with heat resistance and hydrolysis resistance

Figure 4 Mobile phone FOMA N701i ECO (NTT DoKoMo/NEC)

in order to design flame-retardant PLA as a part of business machine (Fuji Xerox, ApeosPort-III and DocuCentre-III series) or high impact strength PLA as a digital solar health meter (TANITA, Ecoliving) [5].

5. 5. 4 Thermoforming

Modified PLA resins for both solid and foamed sheets for thermoforming have been developed to provide thermoformed articles with heat resistance through crystallization during the molding process [5,15]. Potential applications of the thermoforming include food trays, dishes, containers, and cups with heat resistance capable of hot water filling and microwave reheating.

5. 5. 5 Foaming molding

A conventional PLA always shows poor melt elasticity, which is a serious problem for processing operations like foaming molding. By means of material designing of PLA to enhance strain-hardening in the elongational viscosity, a novel PLA resin for foam extrusion process has been developed to provide both foamed sheets for thermoforming and pre-foamed beads for foaming in mold [15]. These foaming articles are also completed with good heat resistance, for instance

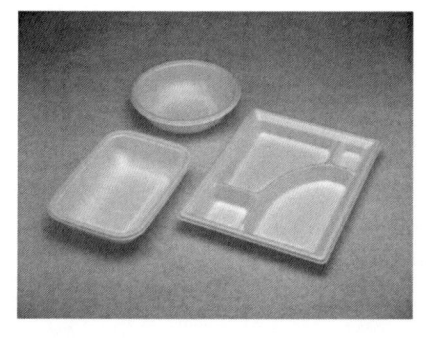

Figure 5 One way use of foamed food trays/containers with heat resistance

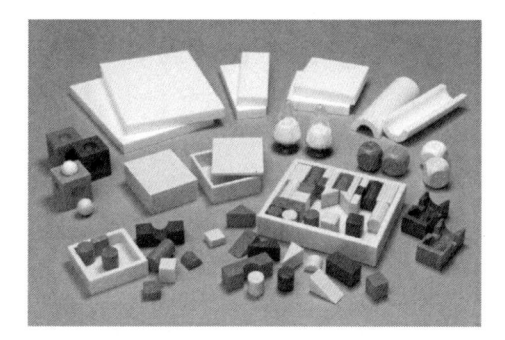

Figure 6 PLA foamed articles with heat resistance (SEKISUI PLASTICS)

foamed food containers (Figure 5) molded by thermoforming being capable of hot water filling and microwave reheating. Also foamed articles (Figure 6) through pre-foamed beads have better dimensional stability than those of expanded polystyrene (EPS) and expanded polypropylene (EPP) at elevated temperature around 150°C.

References

1. H. Tsuji, 'Polylactide', *"Biopolymers, Vol.4 Polyesters III"*, A. Steinbuchel and Y. Doi, *ed.*, WILEY-VCH Verlag GmbH, Germany, 129–177pp (2002)
2. P. Gruber and M. O'Brien, *"Biopolymers, Vol.4 Polyesters III"*, A. Steinbuchel and Y. Doi, *ed.*, WILEY-VCH Verlag GmbH, Germany, 235–250pp (2002)
3. M. Mochizuki, *"Biopolymers, Vol.4 Polyesters III"*, A. Steinbuchel and Y. Doi, *ed.*, WILEY-VCH Verlag GmbH, Germany, 1–23pp (2002)
4. K. Ueda, N. Fukawa, H. Nishimura and M. Mochizuki, *Plastics*, **55**(11), 66–70pp (2004)
5. M. Mochizuki, "Technologies for High-performance and Recycling of Bioplastics", NTS Inc., Japan, 1–60pp (2008)
6. J. Lunt, *Polymer Degradation and Stability*, **59**, 145–153pp (1998)
7. M. Mochizuki and M. Hirami, *Polymers for Advanced Technologies*, **8**, 203–209pp (1997)
8. M. Mochizuki and M. Hirami, "Polymers and Other Advanced Materials", P. N. Prasad, E. Mark and T. J. Fai *ed.*, Plenum Press, New York, 589–596pp (1997)
9. Erwin T.H. Vink, Karl R. Rabago, David A. Glassner, Patrick R. Gruber, *Polymer Degradation and Stability*, **80**, 403 (2003)
10. M. Mochizuki, *SANGYO TO KANKYO*, **33**(3), 83 (2004)
11. J. I. Lauritzen, J. D. Hoffman, *J. Res. N. B. S.*, **64A**, 75 (1960)
12. M. Mochizuki, *Sen'I Gakkaishi*, **66**(2), 70–77pp (2010)
13. M. Mochizuki, H. Shirai, "Advanced Materials and Technologies of Bioplastics", M. Mochizuki and K. Ohshima, *ed.*, CMC Publishing Co., Ltd, Japan, 247–266pp (2009)
14. M. Mochizuki, "Handbook of fibre Structure", J. Hearle, S. Eichhorn, M. Jaffe, and T. Kikutani *ed.*, Woodhead Publishing Limited, Cambridge, 257–275pp (2009)
15. K. Ueda, M. Sakai, S. Murase and M. Mochizuki, *Plastics*, **56**(11), 67–71pp (2005)

Bio-Based Polymers

2013年4月1日　第1刷発行

監　　修	木村良晴	（O0008）
発 行 者	辻　賢司	
発 行 所	株式会社シーエムシー出版	
	東京都千代田区内神田 1-13-1	
	電話 03 (3293) 2061	
	大阪市中央区内平野町 1-3-12	
	電話 06 (4794) 8234	
	http://www.cmcbooks.co.jp/	
編集担当	門脇孝子／倉田恵実	

〔印刷　株式会社遊文舎〕　　　　　　　　　　　©Y. Kimura, 2013

落丁・乱丁本はお取替えいたします。

本書の内容の一部あるいは全部を無断で複写（コピー）することは，法律で認められた場合を除き，著作者および出版社の権利の侵害になります。

ISBN978-4-7813-0271-3　C3058　¥30000E